IntelCenter Words of Abu Yahya al-Libi Vol. 1

ISBN: 978-1606760253

IntelCenter Contact Information
Email: info@intelcenter.com
Web: http://www.intelcenter.com

Phone: 703-370-2962
Fax: 703-370-1571

Mailing Address
PO Box 22572
Alexandria, VA 22304-9257
USA

Table of Contents

Introduction

The words of Abu Yahya al-Libi provide important insights into al-Qaeda. They are essential in understanding its intentions, shifts in focus, difficulties, current and long-term objectives, targeting preferences and more. This volume is designed to provide the intelligence, military and law enforcement communities, as well as researchers, scholars and others, a professional-level reference work bringing all of Abu Yahya al-Libi's primary statements together in one place. It focuses on audio and video statements where al-Libi was the primary or sole speaker. Al-Libi also gave other significant statements during this period in videos where other speakers appeared. Those and other statements will be in Volume 2.

This volume works with three primary types of material. The two most heavily weighted in the selection process are English language transcripts produced by al-Qaeda and exact transcripts of the English subtitling in al-Qaeda's videos. Wherever possible these two types are used. If they are not available, then the best quality translation available of the spoken Arabic was used. At the top of each statement the source material type is indicated in the version section.

Due to the variety of source material, there are noticeable variations in spellings, names and other areas from one statement to another. This was unavoidable as we felt it was critical to preserve the exact copy of al-Qaeda's actual English writings, to include grammatical and spelling errors. This allows analysts and others to use these transcripts in a variety of analytical efforts that would not be possible had everything been standardized.

Sheikh Abu Yahya al-Libi, An Interview with as-Sahab

Release Date: 17 Jun. 2006
Production Date: Unknown
Type: Video statement
Version: Subtitles

Interviewer: All praise is due to Allah, and may prayers and peace be upon the Messenger of Allah Muhammad and upon his family, companions and allies. Honorable Shaykh: if you may, tell us first about the stages of your arrest and how and where it occurred?

al-Libi: All praise is due to Allah, and may prayers and peace be upon the Messenger of Allah and upon his family, companions and allies. After the blessed blows taken by America in the events of September, America announced that the world had been divided into two: either with us or against us. Following this division, the government of Pakistan headed by Pervez Musharraf took a traitorous position as it declared its total support and complete backing of America. Subsequently, the sweeping campaign of arrests began, targeting all Mujahideen residing on Pakistani soil, and not only the Mujahideen but also all foreigners residing in Pakistan, including those legally studying in the religious schools and so on, and others. So during this oppressive, sweeping campaign, many of the Mujahid brothers were arrested, and we received our share of these arrests and I was arrested on May 28th, 2002, in the city of Karachi, at the hands of Pakistani intelligence and police but with the guidance and direction of American intelligence. And as you know, America had declared its all-out war on all Mujahideen, and not merely al-Qaida or Taliban. Its campaign was against all Jihadi movements in general and I was affiliated with a Jihadi group, the well-known Fighting Islamic Group [of Libya] and as part of this campaign, I was arrested and taken to one of the police stations in Pakistan, and after only six hours, I was handed over to the Americans, who were running a prison in Karachi. And that's how I was arrested.

Interviewer: You passed through a number of prisons. Could you tell us about some of those American detention centers?

al-Libi: The story of the prisons is a tragic one. If truth be told, our arrest revealed to us many things which were previously hidden and which one can only discover by experience and by direct contact with the Americans. We found a great difference and huge discrepancy between what we see and hear in the media and the reality which is hidden from many Muslims and even many Mujahideen, due to the blackout America imposes on all media. First of all, I say that the whole world is now part of the United States of America. Its center is Washington and there isn't a country on earth - however much it claims to be hostile or opposed to America - which doesn't have a prison which the US uses against the Mujahideen. In Pakistan–Pakistan is one of the biggest detention centers for the Mujahideen. In Karachi, this metropolis in which there are more than 20,000,000 Muslims, there are American-run prisons. And the country which really contains large prisons and massive detention centers for the Mujahideen and their supporters is Afghanistan. All of Afghanistan has become a prison for the Mujahideen, and the biggest of the prisons used by the US are (1) the central prison, or - as the Mujahideen refer to it - the Prison of Darkness, or the Prison of Torture, which is located in Kabul, and (2) Bagram prison, which is well-known to the media, and from which Allah saved us by His Grace. There's also Guantanamo, similarly well-known. And as for the prisons which are located in the Arab countries, they are countless. When I was arrested in late May, immediately in the very first session, the Lebanese interrogator of course wanted to frighten and terrify me, and so among the threats he made, he asked me, "Which prison do you want us to take you to? To Syria? To Jordan? To Egypt? To Israel?" And then he said, "Even to Libya!" And of course at that time, the relations between Libya and America were ostensibly bad. He told me, "Don't be fooled by what you see in the media about there being some hostility between Libya and America." I would say that in fact all the Arab countries are part of the US and complete the number of US states.

Interviewer: The media have carried pictures showing American troops torturing detainees, especially in Abu Ghrayb prison. If you may, tell us what you saw or heard of violations inside American prisons, especially those that you passed through.

al-Libi: I say that what the media have carried and especially what occurred at Abu Ghrayb in Iraq is insignificant compared to what happens to the Mujahid brothers in the American prisons. First of all, methods of torture are unlimited. That is, the primary goal of the interrogators is extracting information, and their hands are free when it comes to the way they extract this information - i.e., they stop at nothing. And everything you could possibly imagine has been suffered by the Mujahideen. The worst thing we could possibly mention in this regard is violation of honor. Many brothers have been subjected to this thing, and this is not a mere allegation we make, but something we have heard directly from those who suffered this problem. Also, the severe beatings which these prison guards inflict – and especially those brothers who were transferred to the Arab prisons like those in Jordan, Egypt and elsewhere: in these prisons, what the Mujahideen suffer is many times greater than what they suffer in the other prisons run by the Americans directly, like Bagram. In spite of the harshness which they suffer in the latter, if we compare the prisons which the Arab puppet governments control with what our brothers the prisoners and Mujahideen suffer in the prisons controlled by the Americans, we see a big difference, although conditions in all the prisons are severe. There is a special prison for torture in Kabul which is the primary central prison used by the Americans against the brothers. This prison was visited by all the imprisoned brothers - especially the Arabs, non-Afghans, and some leaders of the Taliban and Hekmetyar's Hizb-i-Islami. These all passed through this prison, which consists of solitary (one-man) cells, in front of each one which is a large loudspeaker. Music is played in that prison 24 hours a day, and there are some of the brothers who had to listen to this music - Western, Eastern, and otherwise - continuously for one full year, until they developed psychological complexes, and so you find that some of the brothers, if they hear the least sound of music, they start crying because they are reminded of the tragedy they experienced. Additionally, some brothers had cold water used against them in the biting winter: a barrel full of ice-cold water is brought and the brother - naked and stripped of his clothes - screams "O Allah, O Allah" desiring some mercy, and the arrogant infidel fighter against Allah and His Messenger replies and says to him, "Where's Allah so He can come and remove you from this barrel?" Then the brother is taken and thrown into the cold cell – the brothers remain suspended on the wall at a height of approximately 60 centimeters for six or four or three months – his hands are not freed neither for sleep nor eating nor relieving himself, but only when he is taken for interrogation – the use of dogs to frighten is common, especially in Bagram prison. As for the prison which was established in the headquarters of the Islamic Emirate in the house of Amir-ul-Mumineen Mullah Muhammad Umar, may Allah preserve him: not many know anything about this prison, but it's one of the harshest prisons used by the Americans against the Taliban. The room inside the prison is only one meter by one meter Inside it is a large spotlight which is directed upon the naked brother prisoner, and when the spotlight is turned on, the room heats up because the spotlight is very hot, and in this heat, cold water is all of a sudden poured on the brother. And it continues like this all the time: spotlight/water/spotlight/water, until the brother nearly loses his mind. These are some of the tragedies suffered by our brothers in the prisons, and if we were to continue, I think it would require volumes. This country [America] which claims that it respects the human being and human rights and strives for equality and the lifting of oppression from the Arab peoples and the spread of democracy: this is its real face, about which many Muslims know nothing. And by the grace of Allah, what we saw in their prisons revealed to us their evil and introduced us to their true nature, and we recognized that they are the enemies of Allah and His Messenger. The slogans they raise - some of which we just mentioned - are all phony, and if they ever want to apply them, then on non-Muslims; as for the Muslims, they have no right to them.

Interviewer: Were there any sisters under arrest?

al-Libi: Yes. In the house in which I was arrested there was a Pakistani family living with me on the upper floor, and the woman was arrested by the lowly, despicable Pakistani intelligence, and remained one entire month in prison in another city - not her city. Also, in Bagram prison, from which Allah made easy our deliverance, there was a Pakistani woman who spent two full years in this prison in which there are over 500 prisoners, all of them men of course, except for this one woman in solitary confinement. This woman was, I think, over 40 years old, and was treated in prison like a man in all respects: in her going out for relieving herself, in her being restrained with chains, in her orange clothing with which they clothe the prisoners, in their interrogation of her - i.e., all the techniques they use against the male prisoners were used against this woman. This woman remained 2 full years in prison, until she lost her mind. I still wish that I knew where this woman is, I wish I knew her name, because when we were in prison, we really felt her tragedy and we felt the oppression and suppression she was suffering. We stayed in solitary confinement for 6 months and one month and 2 months - i.e., intermittently, and we know what solitary confinement means, and what those single cells mean. It means being cut off from the world, it means being put in the grave while you're still alive. That's solitary confinement. So because of what we saw of her suffering and oppression, Allah made the way easy for us and we took a stance to help and aid this woman with what we possessed and were capable of. We were captive prisoners and the prisoner is like a slave: he is powerless to help himself. So I and some of my brothers - including some of those who were delivered with me from Bagram prison - staged a hunger strike for 6 straight days. The soldiers came to us and asked us, "Why are you refusing food?" We told them that it was because of this poor woman. Then they said to us that this woman is a criminal and deserves this. But we persisted and were punished and put in solitary confinement for one full month. And all praise is due to Allah: we persevered and Allah bestowed His favor upon this woman and delivered her, This first of all was by the Grace of Allah Alone and then by virtue of the Muslim coming to the aid of a Muslim, and in this is a simple lesson for all Muslims: that you shouldn't belittle anything which you are able to provide to your captive brethren and Mujahid brothers. If Allah the Exalted sees in you honesty, sincerity, and real concern for Allah's religion and the suffering of your brothers, then Allah will put His blessings in this simple reason. You only have this thing which you possess and you can't have more than it. So Allah spited those arrogant ones and responded to this request. And it is from His generosity and favor that He freed us in a way so fantastic that even we, who experienced and lived it, almost can't believe how we exited that fortified fortress of which they are so proud - Bagram prison.

Interviewer: You went through numerous stages in interrogation.

Could you give us a picture of the methods of interrogation followed in the American detention centers?

al-Libi: The truth is, our meetings with the interrogators introduced us to a number of things. First, we used to read in the books that the unbelief of the apostate is stronger and harsher than that of the original unbeliever, and that is a fact which we experienced, and I think the brothers imprisoned in the prisons of the Arab idol-king puppet rulers know this truth. The evilest and most hateful of the interrogators and the most abusive of them to the prisoners are those of Arab origin, especially the Egyptians, Jordanians, and Lebanese. They comprise the most vicious and wicked group among the interrogators whom we saw, whether it be in the dirtiness of the language they use against the prisoners during the investigation, or in the torture and their carrying it out and supervising it by themselves during interrogation and imprisonment. This is a fact which we discovered through experience and not from mere stories or words we've read in books. On the other hand, we discovered that the interrogators, whether they belong to the CIA, FBI, or military intelligence, we discovered the shallowness of their information, whether it be intelligence-related or just their general level of education. I'll give you an example, and it's really an amusing one - let me tell it to you. In the first interrogation session, a Lebanese interrogator and with him an American man sat in front of me. This Lebanese interrogator - who was a Christian - wanted to show me that he understood something of the religion, since I was an observant [Muslim] and belonged to an Islamic group. He wanted to jump on this bandwagon, so he said to me, "You must use your reasoning and be open-minded if you want to save yourself from your present situation." Then he said to me - in a Lebanese accent which I try to imitate - "What does Allah say in the Holy Quran?" I said, "What does he say?" He replied, "Allah says, "I'qilha [tie it] and go ahead." [Prophetic Hadeth] After that, I wanted to find out the meaning of this "verse" which he mentioned according to this "Quran" of his which he knows so well and its connection to his advice to me to use my reason. And the answer came in agreement with the "verse" which he mentioned and he said, "'I'qilha means use your reasoning." So I said to myself, "Abu Hanifah can relax!" This is the level of the interrogators with whom we dealt. I don't deny that among them are experts and devious ones who are skilled in interrogation and have the ability to extract information through cunning and deception, but the majority of those whom we saw and dealt with and with whom our brothers dealt were as we have noted.

Interviewer: By virtue of your arrest, what did you discover about the character of the Americans?

al-Libi: The truth is, we found the American character, or the American soldier with whom we had long-term contact, to be a mix of doctrinal, behavioral, moral, and ideological deviation. I have not found a description more precise and fitting than His statement, Exalted be He: "And those who reject Allah enjoy [this world] and eat as cattle eat; and the Fire will be their abode." (Muhammad 12) And His statement: "But whosoever turns away from My Message, verily for him is a miserable life, and We shall resurrect him blind on the Day of Judgment." (Ta Ha 124) We saw them in comfort and ease, with everything they want in front of them. At Bagram base, the guard sits in front of you guarding the cage with a computer in front of him on which he surfs the Internet. This is the degree of opulence and luxury which they have reached, yet they live lost and alienated. We - Arabs and Muslims - each one of us is proud that he belongs to a specific family and that he knows his father and mother and tribe. This is from the fundamentals of our Arab and Islamic affiliation. But when you come to the American soldier, you find him lost, and estranged. Ask him where his father is, and he will tell you, "I don't know." Ask him what his father's name is, and he will tell you, "I don't know." Just like that, and with utter frankness. And not just one soldier or two or three, but a large slice of the American army. The American soldier will come to you and express to you his social grievance and ruin and he will curse and abuse his mother, saying, "My mother left me when I was little – I'm ruined – I'm destroyed." He says this, but he can't talk about his complaint with his brother soldier because he has the same problem. So he comes to you because he knows that you will listen to him and he trusts you, so he relates his problem to you. The American character is cowardly: one shout from the prisoners shakes the prison and a general alert is declared in the prison. The American character is lost, in search of the way, in search of the path: "Where are we going? We don't know."

The American character engages in purely material dealings: not merely between them and the prisoners, but among themselves also. I mean that the animosity and feuding which occurs between the soldiers is a characteristic inseparable from them. I'll give you an example: in the prison, they are divided into two squads, a squad which works at night and a squad which works in the day. Each squad has a 12-hour shift. The night squad curses the day squad & the day squad curses the night squad. I thought of the statement of Allah: "Every time a new people enters [the fire], it curses its sister." (Al-'Araaf 38) So this is in this world, and it's what awaits them in the hereafter as well. Whenever a problem occurs, they say, "It's all because of the night squad." And the night squad says, "It's all because of the day squad." Also, the American soldier doesn't know the destiny or outcome of this battle which he's gotten into, which afflicts him with distress and complaints, as he awaits the moment in which his tour of duty in Afghanistan ends so he can return to his country. And when the date of his return to America nears, you'll find him in a state of happiness and cheerfulness and exultation, and he'll come and inform the prisoners, saying, "I'll be leaving soon." And some of the prisoners speak with them about conditions in the prison and the hardship they face, and they respond by saying, "We're prisoners like you, we've lost our minds." I swear by Allah, this is what they say! One of them will tell you, "I'm insane, I've lost my mind. I leave the house to go to the prison, and I leave the prison to go to the house, and I leave the house to go to the prison, and I'm here on this base which I am unable to leave." For one full year, and he has one month in which he satisfies his lusts in the countries of his choice.

Interviewer: That's how life was for the prison guards and interrogators. How was it in regard to the Mujahid brothers?

al-Libi: We spoke just now about the condition of the American soldier, and usually, someone who lives free and has all the goods and delights of life provided to him lives in contentment, happiness, and tranquility, while someone who lives in a locked room, forbidden from speaking, with a daily routine and set program for 2 or 3 years will suffer from distress, loneliness and other complaints. However, the reality is that the light of faith contrasts with the darkness of unbelief. I'm talking about the high spirits and relaxation which we experienced and saw in our brothers. And I don't mean just our high-ranking Mujahid brothers, but rather.the general population of prisoners, including the illiterate ones who can neither read nor write. You see all of them in a state of cheerfulness, happiness, and rest. The American soldier will come in amazement and ask, "How can you live like this?" I'll tell you a story: we, the Arabs, were in one room, and when they would take all of us out - and this was related to us by one of the American soldiers - when we would all go out for exercise or showers, he said, "I come and sit where the prisoner sits and sleep where he sleeps, and I feel as if I will explode - I feel like I am suffocating. And I say to myself, how can this prisoner spend all his time in this condition?" The truth is, the state of mind experienced by the prisoners isn't just in Bagram. We passed through many prisons: I alone passed through four prisons and was held in solitary cells, and the brothers we lived with passed through numerous prisons, not only in Afghanistan but also in Egypt and Jordan, until they were brought here. And yet you find them in a state of happiness, laughter, and joy. And perhaps - I swear by Allah that this isn't an exaggeration - perhaps in more of a state of faith, devotion and happiness than they would have been had they been outside the prison. The condition of the prisoners in general is one of high morale, and I state this from experience. Many of the brothers suffered unimaginably awful, difficult conditions, and despite that I have yet to hear that one of the Mujahid brothers went back on his principles or beliefs. On the contrary, there some who may have had some deviation - maybe ideological deviation or a [mistaken] conception of the ground realities - but after living in the prison and coming into contact with these infidels and coming into contact with the Mujahideen, they amended their ideas. And some of them said, "If I had know the Americans and the American soldier had this disposition and were in this state, I swear by Allah, they wouldn't have captured me." And the fact is, the person says this, but his capture was decree of Allah, the Exalted. As for the relationship between the Arabs and the Afghans: it is one of brotherhood, loyalty, and the bonds forged by common faith and creed. The fact is, the Afghans - and the Taliban in particular - don't feel in the least that what happened might be - as is broached - because of the Arabs. They feel that their problem is the same as ours, and also that their destiny is the same as ours. So the relationship was strong, to the extent that when we would be moved from room to room, they would - I swear by Allah - bid us farewell with sobs and weeping, and when we would enter a room, they would greet us by crying. This shows that the bonds of faith are superior to all bonds, as Allah, Exalted is He, said: "Verily, the believers are brothers."

Interviewer: Did you meet any of the leaders of the Taliban, and how were they while in prison?

al-Libi: Yes, we met some of the leaders of the Taliban, and I might not be able to mention some of the names here lest it lead to some problems for them. But I swear by Allah and testify for His sake that the brother Taliban whom we met in prison were among the best we have seen of Allah's worshippers - whether in terms of piety and fear of and devotion to Him or in terms of their disassociation from and rejection of the unbelievers and their methodologies or in terms of their loyalty to the believers, to the extent that they feel that they have been neglectful of your rights. He feels that what happened to you - and you're now together with him in prison - is his fault and that he didn't protect you sufficiently. So those Talib leaders whom we saw were of the Highest standard and truly deserves to be leaders of the Ummah, and I don't say this out of exaggeration, [empty] praise, and excessive flattery which oversteps the bounds, but I say it as testimony for Allah's sake about which I will be questioned in His presence.

Interviewer: What information and news about the outside world used to reach you?

al-Libi: Prison, in reality, means the grave. The prisons which we passed through were tantamount to graves, and the news which leaks through to the prisoners comes from limited sources, and most of it comes from the new detainees who are brought to Bagram prison. They have some information, but because they perhaps spent some time in other prisons, their news is old by the time it reaches us. There was a magazine called "Sulh" ["Peace"] which was published in English, Pushto, and Persian. However, it didn't contain news, but instead, flattery, praise and ideological corruption meant to glorify and convert to the puppet Karzai regime. Sometimes some of the soldiers would come and relate to us some of the news from Iraq, because they felt bitterness - and these are their words - they feel bitter because of their going into Iraq. They say - and this is from their mouths - "we understand our entering Afghanistan, because al-Qaida and Taliban are there, but what is the motive for our going to Iraq? Where are the nuclear weapons and weapons of mass destruction which Bush claimed were present in Iraq? Saddam has been captured, so why are we still in Iraq? Every day, American troops are killed in Iraq, and we - the troops - are the only ones paying the price." So in a demonstration of the bitterness which they feel in their hearts and in a condemnation if their government's blind policies, they come and get it off their chest with stories and news of what is happening in Iraq and Afghanistan.

Interviewer: During your escape from prison, you passed through a number of regions inside Afghanistan. How were you treated by the general public?

al-Libi: We spent a good amount of time, traveled long distances, and visited many houses during our journey which began with our exit from Bagram prison and ended with our reaching the Mujahid brothers. By Allah, every single man whom we met and every single house which we entered was sympathetic to us and supportive of us. By Allah's grace, they sheltered and clothed us and gave us the money that we needed and showed us the way and warned us about the checkpoints which are on the main roads. They gave us a hero's welcome. Our news was well-known to the Afghans whom we met, so as soon as we would show up, they would ask, "You're the four who escaped from Bagram?" And we would reply, "Yes, we're the four who escaped from Bagram." And despite their fear and poverty, they - by Allah - helped us with everything they could possibly help us with, to the extent that some of them - I swear - took off the clothes they were wearing and dressed us in them. So we found them to be completely sympathetic, and we found in them nothing but total hatred for the American forces and their puppet government, the Karzai regime, and disassociation from it and that they are awaiting the day of relief which will come at the hands of the Mujahideen. So what the media have reported - that the Afghan people are supportive of the Karzai regime and that they have acquired stability and economic growth, and that they were happy to put the period if Taliban rule behind them - all of this has no basis whatsoever on the ground, and this is what we experienced and saw, and were it not so, then how could we reach this place if Allah hadn't placed these good-hearted, supportive, helpful at our service? How could we have traveled this long distance? How could we eat and drink? We left Bagram prison without clothes, wearing only pants. How could we have passed through the cities and villages, and traversed the crops and fields? From where did we get our clothing? From where did we get the money with which we bought things? All of this was through the support of these people. When we would enter [a house], even the women would welcome us - event he old women would welcome us and would want to shake our hands and kiss our hands out of love and support for the Mujahideen. And when these Afghans would bid us farewell from their homes, they bade us farewell with tears, sobs, embraces, and fear that we might fall into the hands of [the enemy], and with continuous prayer for us.

Interviewer: While in prison, perhaps you saw some of the leaders of the Mujahideen. Can you tell us about some of them?

al-Libi: Without a doubt - and this is something we must recognize - the Mujahideen have paid a price in the campaign which the US has waged against them. But this is not a shame or defect - we don't say that it is a shame or defect. We call ourselves Mujahideen, and they are those who raise the slogan of Jihad, and what is Jihad? Jihad is made up of exertion, difficulty, exhaustion, and hardship, and in this, they take pride and find prestige. So this campaign involved some of the heads and leaders of the Mujahideen and some of their role-models and some who sacrificed themselves, and their time for the sake of championing Allah's religion. We mention as an example the Mujahid hero Khalid Shaykh Muhammad. The Ummah (Islamic nation) doesn't appreciate the importance of this man, nor the services which he rendered to Allah's religion. And how sorry we were that this brother fell into the hands of the Americans, but Allah - Exalted is He - wanted something, and there's no doubt that what Allah chose for him and the Mujahideen will be better for them. Among them aso is the brave leader Ibn Shaykh al-Libi. I saw Ibn Shaykh al-Libi and we were together for four months in Panjsher prison. I would meet and speak with him and I would ask him, "Shaykh, how's morale?" and he would reply, "10 over 10!" meaning high, even though he suffered severely in the American Prison, and he was a thin, weak man and he's become, as we say, "skin and bones" i.e. very weak, yet he's patient and looks forward to Allah's reward and anticipates the relief of Allah, the Exalted, and he used to say, "If Allah the Exalted wants something, he prepares its causes, and the only reason Allah steered the Americans to Afghanistan - the graveyard of their predecessors the British and the Russians - and to Iraq - whose people are renowned for their fighting, perseverance, and fierceness - is that Allah wants to bring them to an end." That's what he would say. So we ask Allah to release them and the other leaders. And among the leaders of the Taliban, there was Malawi Noor Jalal, the deputy chief of intelligence, a good man who was brutally tortured in the Prison of Darkness in Kabul before being transferred to Bagram, where he was when we left the prison.

Interviewer: Did you have anything to say to the Muslim scholars in particular and the Islamic nation in general?

al-Libi: I say to the Muslim Ulema: Muslim Ulema, what is it you're waiting for? What is it that makes you refrain? I say this to some of the Muslim Ulema who have disowned the Mujahideen, repudiated their actions, and dedicated their pens, pulpits and mouths to slandering Mujahideen. I say to them: don't you know that one day you shall stand in front of Allah? Don't you know that you shall be questioned about every word you say? Don't you know that you will be held accountable for each testimony you give - whether for the unbelievers or against the Mujahideen? Don't you know that this world is short-lived and will pass and end, after which you will harvest the fruits of what you do today? Muslim Ulema: who will awaken the Ummah from its coma? Who will arouse concern in this Ummah? Why do we always hear from the Mujahid scholars, "Go and perform Jihad; Jihad is obligatory in Iraq"? Why don't we ever hear a Mujahid scholar say, "Come to Jihad"? Why don't we hear them say, "Come on, come to us"? Why aren't there scholars in the arenas of Jihad? If Jihad is an individual obligation, is it obligatory on the youth only? What has exempted you? Your knowledge? What has excused you from this duty? Muslim Ulema: it is essential that you free yourselves from this painful reality. You must repudiate these puppet governments which terrorize and frighten you. I swear by Allah, You shall never taste the sweetness of faith, nor the dignity of the believer, nor the glory and power of true belief and certainty in Allah, until and unless you enter the arenas of Jihad and experience the Jihad firsthand instead of from a distance. So we request the Muslim Ulema to stand beside their brothers and not confront them nor be a burden upon them, nor force the Mujahideen to sacrifice some of their energy and effort to respond to their misconceptions. We expect the Muslim Ulema to themselves be provides of fatwas, guidance, education and motivation. Regarding the duty of motivation, this great duty which Allah assigned to His Prophet, saying: "So fight in Allah's Cause - you are held responsible only for yourself - and rouse the believers." (An-Nisa 84): if the righteous, sincere Ulema don't carry out this forgotten act of worship, who will? Who do we expect to say to the Mujahideen, "Go ahead"? Who do we expect to say to them, "Make sacrifices"? Who do we expect to say to them, "Slay the enemies of God"? Why do we always - or almost always - find many of the Ulema standing as an obstacle in the path of Jihad? Is there a clearer and purer banner than the one that the Mujahideen have raised in this era, whether in Afghanistan or Iraq or Palestine or elsewhere? If the Mujahideen have made some mistakes, then that's because of your negligence and absence from their midst. They're making every effort for their actions to be in conformity with the tradition of the Prophet, peace be upon him, and mistakes befall them because they're human or because of their lack of knowledge due to your absence from the field. So if you refrain, it's not the duty of the Mujahideen to join you in refraining and abandon the fields of Jihad and let the enemies of God kill, slaughter, violate honor, and demolish mosques, for us to say, "Leave the Jihad, the Jihad has brought nothing but corruption, destruction, and calamities upon the Ummah." The Jihad which has exposed these puppet governments which have raised the banner of total loyalty to the enemies of Allah, the Jews and Christians: were it not for this Jihad, they wouldn't have been exposed. And today, we see major conferences which are being held why? To combat terrorism. And we see proposals like the proposal made by the Saudi idol-King Abdullah for the creation of an entire organization modeled after the United Nations dedicated to countering terrorism. Why have these proposals emerged at this time? Because of the agony and distress they have suffered at the hands of the Mujahideen. Yes, the Mujahideen have suffered, but this is Jihad. Among them are those who have been killed, captured, displaced, and wounded – some of them are poor, some of them cannot find a place to live or settle, but this is Jihad. You task, Ulema of Islam, your assignment, is today greater than it was before. Today the Ummah needs you, it needs you to stand by it. The Mujahideen are calling out to you and asking you to stand at their side, to enter with them the arenas of Jihad. The Mujahid is waiting to find a sincere scholar who will stand with him in battle. As for the Muslims in general, what we ask of them is what we are doing now. What we ask of all Muslims is that they know that the battle which the Mujahideen are waging isn't, as the media likes to portray it, a battle exclusive to a particular group - a group of bandits or terrorists. No, it's a decisive, crucial battle in which the Ummah plays a part, because the Ummah is being targeted in this battle: targeted in its beliefs, targeted in its morals, targeted in its behavior, targeted in its ideas. So today, the Ummah in its entirety, including men, women and girls, must stand beside its sons. So we tell the Islamic Ummah: a little patience, a little encouragement, and a little sacrifice and support for your brothers the Mujahideen, that we might pick this fruit which for so long blood has been spilled and limbs severed in order to reach it.

[Interviewer asks al-Libi a question in Arabic that is not presented in English subtitles]

al-Libi: One of the greatest of the favors which Allah has bestowed upon us is that he has delivered us from the grip of the wrongdoers. This is a favor. whenever we think about it, we feel so humbled toward Allah. Exalted is He. No one ever thought - no, to this very day, we don't consider ourselves deserving of Allah favoring us in this way. And as we said before, Allah, the Exalted, made things easy for us from start to finish, and brought us to the lands of honor and sacrifice: the lands of Jihad for which our hearts longed when we were in those dark and secluded rooms, and here we are, standing with our brothers the Mujahideen. And we were aid and reinforcement for them, by the grace of Allah, and glad tidings for our brothers: It's an important lesson which Allah has taught us, the Mujahideen, and the Islamic Ummah, as well as the unbelievers whose hearts are full of darkness: that everything is in Allah's Hands, and that if He wants something, He says to it, "Be," and it is. It wasn't by our power that we left prison. We left without shirts on our backs. However, Allah wanted to show us the extent of our need for Him and that we are stripped of everything, so He - to humiliate and spite those infidels and demonstrate their weakness - brought us out from their midst. And the fact is, when I saw pictures of Bagram base , I couldn't believe that it was the base from which we exited, because we weren't let outside without our eyes being blindfolded; it was like a big city - so how did we escape? Only by the grace of Allah, the Exalted. Allah, the Exalted, is the One who favored us with our exit from prison, and it's He who made the way easy, & it's He who has sheltered us like this, so with His permission, we shall continue on this path, the path of Jihad in Allah's way. We've tasted the pleasure of Jihad and we can't leave it. And by the grace of Allah, He has shown us the weakness of our enemies and shown us our strength by way of our weakness. When we were weak, Allah showed us that we are strong, so we shall, with Allah's permission, support our brothers, and we are a part of them, and we shall, Allah willing, combat these Christians and their apostate helpers, and we shall never abandon the path of Jihad. Allah willing, we won't leave the path of combat until Allah grants us martyrdom or rules between us and them, and He is the best of rulers.

[Interviewer asks al-Libi a question in Arabic that is not presented in English subtitles]

al-Libi: Yes. I continue to tell the Mujahideen what Allah told them: "O you who believe! Persevere in patience and constancy; vie [with one another] in such perseverance; strengthen each other; and fear Allah; that he may prosper." (Aal Imraan 200) So you must be patient and avoid all sins, for they are the gateway to defeat. We aren't defeated by our lack of manpower and weaponry, but are defeated by the sins we commit and our disobedience to Allah. So we - and this is directed to all Muslims but especially the Mujahideen among them - we must fear Allah and be very humble and break down in front of Him, and avoid conceitedness and self-importance, and always be aware that every victory we win is a Divine gift and present for which we must be thankful. And part of being thankful for it is that we preserve it. And I say to my captive brothers whom we left behind, whether those whom we saw or those whom we did not: we're continuing on the path, with Allah's permission. And I say to the leaders of the Mujahideen, and foremost among them the chivalrous imam who lives outside his time and who renewed for the Ummah the meaning of loyalty and disassociation, the Commander of the Faithful, Mullah Muhammad Umar: to him I say, we are still committed to our pledge and are still on the path, and I give you the good news that Allah's victory is coming your way, and I give you the good news that Allah shall give you power on earth better, stronger, and wider than you had before. And know that what you lost is nothing and that what Allah has given you of the Muslims' love of you and their prayers for you night and day is greater than what you lost. And we say to the renewer of the banner of Jihad in this era, the Mujahid Shaykh Usama bin Ladin: continue with Allah's blessings, for Allah shall make you happy with the demise of this arrogant, immoral nation. Your helpers are behind you, and shall neither stop nor be stopped, with Allah's permission. Also, we say to the unsheathed sword which Allah has drawn against the enemies of Allah and His Messenger, the virtuous Mujahid Shaykh Abu Musab al-Zarqawi: slay the enemies of Allah, and slay them some more. And know that Allah the Exalted is the one who bestowed on you this status. And be conscious of the trust which has been put on your shoulders, and know that the Ummah looks forward to more from you, and that the oppressed and displaced are looking for and awaiting a refuge in which to take refuge. So don't neglect his trust but preserve and protect it. And we say to the virtuous Mujahid commander Shayk Abu al-Layth al-Qasimi: Allah has prepared this matter for you and raised your status because you took the path of Jihad in Allah's way so be firm on this path and don't glance right or left nor become upset or concern yourself with the collapsed and fallen ones who litter the length of the path. And I also address the Mujahid leader and ascetic worshiper whom we consider to be so but we do not purify before Allah, Dr. Ayman al-Zawahiri: know that Allah has made it easy for you to take the path of Jihad and made you firm on it, so keep up your motivation of the Ummah and convey to it your words which move the hearts of the Muslims and enrage the unbelievers, and be firm on this path and increase your guidance, advice and direction to the Mujahideen in the East and West. Also, we can't forget our brothers the captives in the prisons of the apostates, whether in Egypt like the Mujahid Shaykh Abu Yasir Rifai Ahmad Taha, or our brothers in the prisons of Libya like the virtuous Shaykh Abu Munadir al-Saa'idi and the leader Abu Abdullah al-Sadiq. Also, the patient Shaykh in Jordan, Abu Muhammad al-Maqdisi, and the Shaykh of the patient ones in this era, the man who taught the Ummah how the scholar should be, how his patience should be, how his sacrifice should be, how his declaration of the truth should be, and that the Mujahid should declare this word and not fear the blame of a blamer, Shaykh Umar Abd al-Rahman, may Allah free him and those brothers who are with him. I have no doubt that one of the reasons for the continuing destruction and misfortune which has struck America is its animosity to this man. As the Prophet, peace be upon him, said [in a Hadeeth Qudsi], "I have declared war on he who is hostile to a friend of Mine." And so this is our message to the captives and all the Mujahideen in the East and West, and we send our sincere greetings to the martyrdom seekers in Palestine and Iraq, those who have brought back to the Ummah the meaning of sacrifice and giving, those upon whose firmness, strength and determination the boulder of international unbelief led by America was dashed to pieces. And it's my pleasure to dedicate them some words which perhaps don't do justice to their sacrifices, but they are all I'm able to offer here. I shan't lament he who sold all that is worldly And bought the other and looked to eternity And advanced fearlessly and with all his heart His faith gleaming and shining like a spark An igniter of the war uniquely and courageously If it is said "who is its lion?" he says "I am he" Firm of heart and bold of chest His friend is combat and hard work his guest In his soul are concerns if they were to be poured Onto a lofty peak, from their weight it would be felled He did not acquire wealth but spent his years In the shade of the sword or redness of the spears He remembers Allah and cries while The fool laughs in this world of smiles This, and I send peace and prayers upon the best of Allah's creation, Muhammad, and upon his family and companions.

Light and Fire in Elegizing the Martyr Abu Musab al-Zarqawi

Release Date: 30 Jul. 2006
Production Date: Jun. 2006
Type: Video statement
Version: English transcript

Elegizing Shaykh Abu Mus'ab al-Zarqawi may Allah have mercy on him

Shaykh Abu Yahya al-Libi
Jumada al-Uulaa 1427 AH / June 2006 CE
As-Sahab Media

All praise is due to Allah, who dignifies Islam with His aid, humiliates polytheism with His force, directs all affairs with His command, and draws out the infidels with His plotting: the One who preordained the alternation of days with His justice. And may peace and prayers be upon him with whose sword Allah raised the beacon of Islam.

To proceed...Ummah of Islam: peace be upon you and the mercy of Allah and His blessings.

Today, one of the lions of Islam was killed, and one of the heroes of Tawheed was martyred, and with that one of the igniters of the wars passed away: splitter of the heads of the Romans, the sword unsheathed against the enemies of the religion, the Mujahid commander Abu Mus'ab al-Zarqawi, may Allah the Exalted have mercy upon him. And so a bright page of contemporary history was closed, a page whose lines were recorded in ink of blood, and whose words were written in the light of resolve, and whose letters were embellished with the pearls of faith.

Yes, Abu Mus'ab was killed, in the way that men are killed in the battlefields, and he went out like the heroes go out: amidst the blows of the swords. He bade farewell to this world as a steadfast, confident emigrant Mujahid who was neither swayed by the severest of tests nor broken by the distractions of the time. How often he waded into battle without fear, and defied danger without hesitation or doubt. And how much he pursued death at every turn, and knocked at its doors at every rendezvous, and he continued to seek it and strive for it until he earned the honor of dying in the same way that he wore the clothes of glory and chivalry in his life.

He tosses onto the arenas some of his blood To say: O world, come look and testify For here we feed the fields of Jihad By pouring on them waves of fresh blood And here is the believers' Ribaat and a place For their Jihad or a sign for he who is guided

And he is deserving - and so he should be - of the statement of the Prophet, peace be upon him, "One of the best examples for the people to emulate is a man who holds the reins of his steed in the Path of Allah and who, every time he hears a cry for help or the sound of fear, runs to it seeking death where it is to be found." And it is as if I am with him as he roams the fields of Jihad and goes into battle, and eagerly awaits the day of martyrdom, shouting to the skies with his words and actions:

I shall carry my soul in the palm of my hand And toss it into the pits of death So either a life which delights the friend Or a death which enrages the foes

Yes, Shaykh Abu Mus'ab al-Zarqawi was killed, to establish with proof positive that the brilliant words which he would send to his Ummah and which he would pull from the bottom of his heart with honesty and sincerity were not just empty claims, pretty words or flowery sermons, but rather, were a methodology, doctrine, and principle for which blood is spilled, for whose life limbs are severed, and for whose superiority life is given. And he didn't suffice with letting the evidence for his sincerity be his effort, sweat and lack of sleep: no, he made his finale his blood and soul, to say to the whole world, "My faith, religion and methodology are more valuable and important than anything, even my soul which is in my body, and the proof is in what you have seen with your own eyes, and not in what you have heard or been told."

Yes, Shaykh Abu Mus'ab al-Zarqawi was killed, but do his enemies who applauded his death realize that they have presented him with the greatest wish, noblest gift, and highest goal he could ever ask for? It is martyrdom in Allah's path, and is there a wish greater than martyrdom? "Think not of those who are slain in Allah's way as dead. Nay, they live, finding their sustenance from their Lord. They rejoice in the bounty provided by Allah: and they rejoice over those behind them who have not yet joined them, that on them is no fear, nor will they be sad." [3:169-170]

And when they saw that some ways of life are humiliating For them, and that a dignified death is not prohibited They refused to live a life worthy of blame And died a death in which there was no blame And don't be surprised if the lions are vanquished By the enemy dogs among the Arabs and Ajam For the spear of Wahshi made Hamza taste death And the death of Ali was in the sword of Ibn Muljim

Yes, Shaykh Abu Mus'ab al-Zarqawi was killed, but his words have not died, nor has his philosophy been cut short. And how could it, when the love of Jihad, martyrdom, and sacrifice which he sowed in the hearts of the youth has continued to develop, flourish, bloom and bear fruit, until the martyrdom-seekers - youth and elders, men and women - have become row after row whose slogan is:

I shall take revenge, but for my Lord and religion And I shall remain on my path with conviction So either to victory over the people Or to Allah with the immortal ones

And here I say: you truly consoled the Islamic Ummah - oh rogue of Iraq Nouri al-Maliki - when you prefaced the news with your words, "To all the Zarqawians," because you knew that behind al-Zarqawi are Zarqawians who have not died, and whose ideology is his ideology, whose faith is his faith, and whose path is his path. They will not change nor swap, and their way is Jihad, their provision is patience, and their motivation is love for the Gardens.

If Abu Mus'ab has disappeared from us The armies of guidance have not disappeared They shall continue with resolve in his path Cleaving the darkness of the enemy like meteors

So let the worshippers of the cross, dynasties of treason, suckers of depravity, and agents of the Jews who boasted about their participation in the killing of the Shaykh know that their swaggering won't be for long, and that the reprisals of the Mujahideen shall come like lightning bolts one after the other, and that the retaliation for every fragrant drop of the Mujahid commander's blood shall be swift, heavy and costly, and that this boastfulness of yours which pleased your infidel masters shall be a bitter curse upon you which will make you bite your knuckles in regret. So wait: we too are waiting.

So oh Ummah of Islam: here is one of your righteous, loyal sons who was sincere in his affiliation to you, jealousy for you, and eagerness to honor you and raise your standing, and who gave his life in exchange for the superiority of your religion, establishment of your Shari'ah, and liberation of you from the disgrace of slavery and vassalage to men. So be loyal to him by staying on his path and continuing to support his companions and helpers with your self, wealth and prayers. And know that Shaykh Abu Mus'ab, Allah have mercy on him, was not killed in pursuit of material gain, nor in search of the goods of this world, nor out of desire for its adornments, for his soul and concerns were too noble to stoop to the level of this lowly world. He was the one who inhaled the breezes of glory on the fields of battle and truly knew how the believer is honored with his religion and faith, and he wouldn't have exchanged it for anything. And so he traveled along this path with patience and diligence, overcoming difficulties and throwing off shackles until Allah appointed him to that lofty rank.

He continued to strive until those jealous of him said He has a short way to highness

As for you, the sincere, patient Mujahideen: may Allah console you. And you should know, beloved brothers, that this religion is the religion of Allah, and He has taken it upon Himself to protect and preserve it, and the martyrdom of the Mujahid Shaykh is only another part of the series of trials with which Allah tests His believing slaves, to reveal their reliance on Him, confidence in Him, and steadfastness on His religion. "What you suffered on the day the two armies met was with the permission of Allah, in order that He might know the believers and know the hypocrites, who were told, 'Come, fight in the way of Allah, or [at least] defend'. They said, 'If we knew of a fight, we would follow you'. They were that day nearer to unbelief than to faith." [3:166- 167]

So fear Allah in your Jihad, fear Allah in your steadfastness, and fear Allah in the blood of your brothers. And know that the greatest act of loyalty to them - and in fact, there is no loyalty to them without it - is to remain firm on their path and hold fast to the methodology for which they sacrificed and were killed. And avoid side roads, for you have been left with the clarity of the Book and Sunnah, whose night is like day and from which only the irredeemable deviate. "O you who believe! Persevere in patience and constancy; vie [with one another] in such perseverance; strengthen each other; and fear Allah; that ye may prosper." [3:200]

Oh Allah, send prayers on Muhammad and all his companions and family. And may peace be on you and the mercy of Allah and His blessings.

Combat, Not Compromise

Release Date: 2 Nov. 2006
Production Date: Oct. 2006
Type: Video statement
Version: Subtitles

I offer my condolences to myself, my brothers the Mujahideen, and the Islamic Ummah in its entirety, on the loss of a hero of Islam and departure of a knight of Jihad.

He is Farooq al-IraqiÉ

And you, my Mujahid brothers in Iraq, must observe patience and unityÉ

Oh scholars of Islam: if you are remiss, who will lead the way?

Oh scholars of Islam: the battle awaits you, and the fields of Jihad, preparation and strength await you and look forward to you. And by Allah, you will find in them nothing but respect, honor and pride from your devoted sons the Mujahideen.

[Statement begins with English subtitles]

All praise is due to Allah, and may peace and prayers be upon he Messenger of Allah and on his family, companions and allies.

As for what follows: Allah, the Glorious and Great, said,

"And say not to those who are slain in the way of Allah 'dead'. Nay, they are alive, though you perceive not" (2:154).

I offer my condolences to myself, my brothers the Mujahideen, and the Islamic Ummah in its entirety, on the loss of a hero of Islam and departure of a knight of Jihad. He is Farooq al-Iraqi, whom the fields of Jihad came to know as he visited and roamed them and shuttled between them, seeking to help the religion of Allah, the Glorious and Great, striving to honor His Word, and toiling earnestly to aid persecuted and oppressed Muslims everywhere.

Bosnia and Herzegovina came to know him when it was invaded by the atheist, Serb army and he rushed there with his brothers the Mujahideen, to fight, compete and persevere until the matter came to an end and finished in the way it did. But he didn't permit himself to return to the worldly life and wallow in its delights.

Instead, he searched for other fields of Jihad, because his soul had become saturated with love for this act of worship. And thus is Jihad. He who has tasted Jihad cannot leave it, so long as he is sincere with Allah, aware of the importance and status of this act of worship, and knows the virtue of the Mujahideen and the rank which Allah has accorded them. So he went to the Philippines, to those forests and jungles, and spent months there with his Mujahid brothers and their poverty, weakness and displacement. After that, he moved on to Indonesia, not as a tourist in search of the pleasures of this worldly life, but as an advocate of the persecuted, displaced Muslims who were being torn by the knives of Christian hate in the Maluku Islands. He engaged himself in training his brothers there in the use of all the weapons in which he had gained expertise in the land of Jihad, until he fell into the hands of the Christians as a captive in this fierce Crusade which has devoured everything. He spent three years in their prisons before Allah the Glorious and Great favored him with the miracle of escape from Bagram prison, but he refused to return to the life of this world and his soul didn't lean toward its delights and attractions. Instead, he began to search for the fields of Jihad, this act of worship which he adored and was in love with. So he moved on to Iraq, and there Allah decreed for him what he always desired and what every Mujahid Muslim wishes for: martyrdom in Allah's path.

Farooq al-Iraqi, that man whose body was thin and weak, but who was heavy in faith, strong in determination, and tough in his zeal.

That's why the enemies of Allah, the Glorious and Great, only dared to approach the house in which he resided with an entire army trained in the use of the most modern weaponry: the Christian British Army.

But this Mujahid man refused to surrender, after having tasted the bitterness of captivity and repression for three years, and so he fought until he was killed.

We ask Allah to raise his level and house him in the highest Garden of Paradise.

And here I tell my Mujahid brothers: even though the killing of the likes of these champions affects our souls and induces sorrow in our hearts - for they are our beloved brothers and companions on the path - it will not weaken us or cause us to backtrack or leave the arenas of Jihad. And this act of worship, by Allah, only lives with the blood of the likes of these who have offered themselves and their blood, souls, hearts and entire lives as a ransom for Allah's religion.

Farooq al-Iraqi was preceded in the history of Islam by many champions who were killed at the hands of the unbelievers and criminals, yet neither was the torch of Islam extinguished nor were the flames of Jihad put out. Rather, they continued to burn, increasing and gaining in strength, and the tree of Islam and Jihad was irrigated with their blood.

So we tell our Muslim brothers: beware of despair, beware of despondency. Be firm and steadfast, and know that Allah is with you and that Allah is with the patient ones, believers and righteous. And know, oh Mujahid brothers, that the omens of victory have begun to appear everywhere: in Afghanistan, in Iraq, there on the soil of Chechnya, and in Palestine.

And know that the power of this unbelieving enemy, who has filled the earth with his corruption, tyranny and arrogance, has begun to break, and that his strength has began to wane and his wind is dispersing.

Look at what they are suffering in Afghanistan: by Allah, what you hear and what they put out every now and then doesn't equal even one-tenth of what they are actually suffering. They are frightened, terrified, split up and displaced, and the only place they find for themselves are their refuges, which they didn't prepare for fighting and combat, but rather protecting themselves in their military barracks.

Here is the Taliban movement under the leadership of the Commander of the Believers, Mulla Muhammad Umar, may Allah protect him, who is personally leading the battles - and I stress, by Allah, he is personally leading the battles - here is the Taliban movement starting to return to the arena with a strength recognized by its enemies before its friends.

Oh Mujahid brothers in Afghanistan: persevere, and strengthen your resolve, and be severe with your enemies, and know that Allah is with you and that victory is close at hand, after which your state will return to you better, stronger and bigger than before. And you, my Mujahid brothers in Iraq, must observe patience and unity, and must know that the ways of Allah are not partial to anyone. And know that victory comes through agreement and unity and with harmony and brotherliness. And know that defeat and breakup comes with dispute, dissent and discord, as Allah, the Glorious, the Great, said,

"O you who believe! When you meet a force, be firm, and remember Allah much, that you may prosper. And obey Allah and his Messenger, and fall into no disputes, lest you lose heart and your power depart. And persevere, for Allah is with those who persevere" (8:45-46).

So come together, close ranks, strengthen your resolve, and be severe with your enemies.

Oh my brothers the Mujahideen in Iraq: you were a cause of this Ummah being revived after its sleep and awakened after its slumber.

So brothers, beware of wasting the fruit of this Jihad in which the Muslims have offered all they have by sacrificing their blood, wealth, sweat, effort and prayers.

The Jihad in Iraq - like every Jihad - is a trust on the shoulders of those carrying it out, about which they shall be asked before Allah the Exalted. And you are worthy - as we see it - of being leaders of this Ummah and abandoning trivialities and base things, for Allah the Exalted has lifted you to the pinnacle of worship - which is Jihad in Allah's path - so it is not for you descend to the level of selfish desires and claims.

Oh brothers, oh Mujahideen: Allah the Exalted is showing you the miracles of His victory: after hardship, comfort; after difficulty, ease; and after severity, prosperity.

Here is your primary enemy, who filled the earth with his boastfulness, vanity and pride, admitting in all humiliation that his going to Iraq was the wrong decision to make.

We ask him: how and when did you discover that your entrance to Iraq was a mistake?

Was it after your aircraft became sick and tired of transporting thousands of coffins full of the rotten corpses of those killed at the hands of the sincere and perseverant ones there?

Was it after you spent millions, only to find that all that money went for nothing?

Was it after you discovered that the nation of Islam is the nation of faith, nation of sacrifice and nation of strength: the nation which never compromises on its values and doesn't allow its leadership to be handed to others?

And we tell him: be aware that you shall discover that your going to Afghanistan was also a mistake.

Be aware that you shall discover that your backing of the state of Israel was also a mistake.

And be aware that you shall know with certainty that your entrance to the Arabian Peninsula - cradle of the Message and the place where the revelation descended - was a mistake as well.

And you continue to recognize and admit that this nation is the nation of steadfastness, sacrifice and strength which isn't overpowered by its enemies, whatever their strength and regardless of what their despotism and tyranny has come to. And history conveys to you our events, affairs and battles.

To conclude, I direct a word to the scholars of the nation, to the carriers of the inheritance of Prophethood, to those who Allah the Glorious chose to be heirs of his Prophet, peace be upon him, and I say to them:

oh our honorable scholars, oh our great scholars, our brother the Mujahid Shaykh Abu Hamza al-Muhajir has addressed to you a speech which would make boulders melt, by Allah, in which he beseeches and encourages you and introduces you to the conditions of the Mujahideen and the need of the fields of Jihad for those like you.

So we tell you: oh our honorable scholars, the fields of Jihad are awaiting you, and the opportunity has not passed you by, nor will it pass you by. So rush and go forth to them, and be in the vanguard of the caravan and at the front of the convoy, for you are the leaders of the Ummah and its commanders and guardians.

Oh scholars of the nation: what has made you abstain from this great act of worship? What has made you refrain from performing this major obligation about which the Prophet, peace be upon him, said

"By He in whose hands is my soul, were it not that I would create difficulty for my nation, I wouldn't stay behind any brigade raiding in Allah's path."

Oh scholars of Islam: has the love of this worldly life made you refrain from this duty?

We absolve and purify you of that, for we have long heard from you explanations of the statement of Allah, the Glorious and Great,

"O you who believe! What is the matter with you that when you are asked to go forth in the cause of Allah, you cling heavily to the earth? Do you prefer the life of this world to the hereafter? But little is the comfort of this life, as compared with the hereafter" (9:38).

Oh scholars of Islam, what has held you back from the caravan of the Mujahideen: is it the fear of death and being killed? How often we have heard from you recitation and explanation of Allah's statement,

"Wherever you are, death will find you, even if you are in towers built up strong and high!" (4:78)

And how often we have learned from you that when a man's time comes, he will not go past it, and that Allah said,

"Nor can a soul die except by Allah's leave, a term fixed by writing" (3:145).

Oh scholars of Islam: what has held you back from the arenas of Jihad?

Is it the fear of the severity, violence and strength of the enemy? How often have we heard from you recitation and explanation of Allah's statement,

"Those to whom the people said, 'The people are gathering against you, so fear them,' but it [only] increased their faith, and they said, 'For us Allah suffices, and He is the Best Guardian'" (3:173).

And how often have we heard from you recitation and explanation of Allah's statement,

"It is only Satan that suggests to you the fear of his friends: Be not afraid of them, but fear me, if you have faith" (3:175).

Oh scholars of Islam: your nation is writhing in its wounds; your nation is writhing in the hell of the tragedies, hardship and difficulty to which its enemies have driven it. Here is your nation in Afghanistan, Iraq, Chechnya, Palestine, and Kashmir - and moreover, in all Arab countries - where the prisons are full of the best of Allah's saints. Here are the apostate atheists boasting everywhere that they have the upper hand. And here are its apostate enemies who have gained control of it preventing Allah's worshippers from going to Allah's mosques in broad daylight, only permitting them to enter them with admission cards, as if they are not houses of Allah, the Glorious and Great, but rather, the guesthouses of theirs.

So oh scholars of Islam: who will declare the word of truth?

Who will stand up to these criminals?

Who will expose their tyranny and the unbelief, atheism, apostasy and war against Allah's religion of which they are guilty?

Oh scholars of Islam: if you delay, who will step forward?

Oh scholars of Islam: if you are remiss, who will lead the way?

Oh scholars of Islam: the battle awaits you, and the fields of Jihad, preparation and strength await you and look forward to you. And by Allah, you will find in them nothing but respect, honor and pride for your devoted sons the Mujahideen. And you will fight in them nothing but the glory of religion, deliciousness of faith, strength of conviction, victory and domination, and healing for the chests of a believing people.

I ask Allah, the Glorious and Great, to strengthen our hearts and yours, and to make you and us firm on the truth until we meet Him.

And our final prayer is that all praise is due to Allah, Lord of the Worlds.

And the Crusade Continues... The AIDs Children in Libya

Release Date: 1 Feb. 2007
Production Date: Sep. 2005
Type: Video interview
Version: Subtitles

In the name of Allah, the most Compassionate, the most merciful.

All praise is due to Allah and may peace and prayers be upon His chosen Prophet and his family and companions.

As part of the series of crimes which the Crusader West presents from time to time as civilized gifts which fit in with its democracy, slogans and values, an old/new issue has again come to the fore to be a new poisoned dagger stabbing the body of our torn Islamic Ummah. All have heard of that heinous crime for which we haven't found an appropriate description in the dictionary, due to its extreme ugliness and lowliness; that ugly crime which five Bulgarian nurses and a Palestinian doctor committed by injecting more than 400 Libyan children with the AIDS virus in a Benghazi hospital, injections which carried latent hatred and criminality; that crime which indicated a complete abandonment of all human values and revealed hearts in which hatred is embodied in its ugliest form.

So hatred has appeared in its most hideous state, in which murder is committed in the most despicable of ways, for this event to become a prominent title which introduces us to the values, civilization and real face of the West.

It is a crime which wasn't committed by their armies which are used to criminality and have accustomed us to its ugliest forms, but was rather committed by those who traveled thousands of kilometers to treat and care for the people - or so they claimed - but suddenly, the nurses turned into cold-blooded, evil-spirited murderers with hate-filled hearts.

"How [can there be such a covenant], when if they get victory over you, they neither respect in you the ties of kinship nor of covenant? With [fair words from] their mouths they please you, but their hearts are defiant, and most of them are rebellious and wicked." (9:8)

It is a crime whose weapon of destruction wasn't the B-52 or F-16, nor were destructive 7-ton rockets fired, nor was there any dropping of the smart bombs whose smartness always leads to the murder of hundreds of women, children and elderly, for us to be told after every massacre that war cannot be free of mistakes.

No, the deadly weapon in this crime was an injection which should have contained medicine and healing, but instead introduced into the bodies of innocent children the lethal poison which would cause them constant pain and a slow death. It is a crime which didn't take place in the mountains of Tora Bora, nor the streets and houses of Fallujah or Tal Afar, where a war which devours rocks and trees and melts man and stone under the pretext of the presence of "terrorists" in the area, which - for the civilized West - is a sufficient excuse to legitimize the annihilation of entire villages with their inhabitants.

No, the scene of the crime was a hospital to which people go to seek treatment and medicine and to flee from illness, but suddenly, the hospital became an abattoir from which the healthy come out sick and the sick come out dead. It is a crime which was not committed against people possessing weapons with which they are defending themselves, for the lying hypocritical media to tell us that they are suspected extremists plotting what they call terrorist attacks.

No, this war - and I call it a war - was waged in its most hideous form against children who carry innocence and purity their hearts and souls.

This, in brief, is the form the crime took, even though words are truly incapable of property describing it.

Has the world witnessed hatred stronger than their hatred?

Have you heard of a deed uglier and more atrocious than their deed?

Has anyone seen savagery on a par with their savagery?

And did anyone ever imagine that their criminality would descend to such lowliness?

Those who know the truth about these hateful infidels and haven't been deceived by their pretty words and slogans and lying propaganda won't be surprised by such a thing, because the Quran has summarized for us this long list of consecutive crimes in a single verse with which it exposed the truth about them, revealed the secrets of their souls and described their actions with precision: "If they overcome you, they behave to you as enemies, and stretch out their hands and their tongues against you with evil. And they desire that you reject the truth." (60:2)

The infidel Crusader West has accustomed us to constant boasting that theirs' is the best of societies most advanced of civilizations and the one most protective of the rights of man - and animal too - and that the world can only extract itself from its crises by following their path of democracy and doing as they do.

So what has been the stance of their civilization and their democracy towards this ugly crime?

Did you hear any one of their tyrants, or any one of their human rights foundations, or any of their charities, or anyone of their independent organizations merely talk about - much less condemn - these victimized children, even indirectly, and mention their suffering and the suffering of their families or even shed a tear in sympathy and mercy?!

No, instead of that they rushed shamelessly to express their sympathy and support for the murderous wolves as they bit their victims with their teeth and tore them apart with their claws, demonstrated indifference to the victims' bleeding wounds, loud cries and continuous suffering, and directed all their effort into displaying sympathy and mercy for those wolves, for no other reason than that they are wolves like them.

"The unbelievers are protectors, one of another." (8:73)

What business does America have with five obscure Bulgarian nurses which makes its president promise to seek their release - their release and nothing else, not even extradition for trial in their supposedly just legal system?!

They will accept nothing less then their release, so let their crime be, let more than 400 children go to hell and let their families perish in their rage.

And what business does the old hag Britain have with these criminals which makes her dog express his great sorrow and disappointment at the judgment passed on them and promise to do all he can to secure their release?!

It is the Nazarene faith which binds them to each other, and their statements embody the motivations of hostility and hatred to Islam and Muslims. And their hypocrisy, charlatanism and camouflaging didn't help them to take any stance other than this stance, with which they disgraced themselves and introduced themselves to all who didn't know them before.

The case of the AIDS children isn't the first case which the West has treated with such duplicity. No, their application of double standards, especially when the victim is a Muslim, has become the most distinctive feature of their policies, stances and resolutions.

There is Palestine - and what will make you realize what Palestine is - in which the lives of dozens of unarmed men, women and children are taken at the hands of the criminal Jews.

Yet despite all of that, the Crusader West has only given the people of Palestine a lethal embargo and death-defying support to the Zionist murderers.

When one of the soldiers of the state of the sons of Zion is killed, no sooner does the news reach them than they rush to condemn and deplore and send their condolences to the family of the "victim" - yes, "victim" in their dictionary. But when entire territories are bulldozed along with everything in them, and houses are demolished onto their residents, and mosques are bombed with the worshipers inside, and cars are melted along with their passengers, and children's limbs are scattered in the streets, and the cries of the helpless women and children fill the air, none of that pricks the hearts of the infidel Crusader West, nor is it considered in their dictionary to be a crime deserving of condemnation or indignation.

Why? Because those killed are Muslims, and as long as that's the case, they're not victims deserving of sympathy, mercy or the continuous calls to stop the violent cycle of killing. And the same goes for Iraq, Afghanistan, Chechnya, Somalia and other Muslim lands.

This, then, is the West, and these are its values which it wants to offer us, and this is its democracy which it seeks to spread among us.

So away with them and their civilization, values and "justice".

And this crime is a living example of the disregard shown by the regimes of treason and treachery for the lives of their repressed, helpless peoples and their scorn for their dignity. If not, how then were these butchers able to commit this crime against such a large number of innocent children without the regime of the tyrant of Libya, al-Qadhafi, having some knowledge or suspicion about it - or indeed, without colluding and taking part in this crime?

Where was the regime during the period in which the children were being fed the virus of murder and criminality, after which it portrayed itself as eager for justice, in search of the facts and in pursuit of the criminals, even though it is the leader in crime and corruption?

Where were the foundations of this crumbling regime?

Where was its surveillance all this time?

And how is the heinous crime committed in total secrecy and utter silence without it being noticed by its security forces and intelligence agencies which count the peoples' breaths and follow them in all their affairs, whether minor or great?

Or was it busy with pursuing and chasing sincere Muslims to flatter the West and please its masters, just like it will please them by dropping the case of this heinous crime, as you will see in due time?

That's why we say that the time has come, O Muslims, for you to recognize these hateful infidels as they really are. It's time for you to recognize these hateful infidels as they really are. They are murders, killers, and criminals who sow corruption on Earth. And don't be fooled by their shiny slogans and false words, because - by Allah - your blood, lives and children are worthless to them.

You will only find honor with which to make progress and strength with which to protect yourselves in your true religion, which is based on disowning them and their servants and declaring open animosity to them. And their evil and corruption will only be stopped by fighting without leniency and performing non-stop Jihad until there is no unbelief and all religion is for Allah.

"Were it not that Allah repels one set of people by means of another, there would surely have been destroyed monasteries, churches, and synagogues, and mosques in which the name of Allah is mentioned much. Allah will certainly aid those who aid Him, for verily, Allah is strong and mighty." (22:40)

And all praise is due to Allah, first and last.

Iraq: Between Indications of Victory & Conspiratorial Intriques

Release Date: 22 Mar. 2007
Production Date: Sep./Oct. 2005
Type: Audio statement on video
Version: Subtitles

All praise is due to Allah, and may peace and prayers be upon the Messenger of Allah, and upon his family, companions, and allies.

Beloved brother Mujahideen in Iraq of honor and Baghdad of the Caliphate: Peace be upon you and the mercy of Allah and His blessings.

Allah has honored you today in the place in which you are with the greatest thing with which He honors His believing slaves: the worship of Jihad, which you have successfully resurrected from the dead, and with which Allah raised your status, spread your fame, healed your chests, and pleased through you your brothers.

He, Glorious and Exalted is He, has also favored you by giving you a large part to play in shaking the pillars of the modern-day 'Ad and contemporary Hubal for which tyrants East and West would fall to the ground kneeling and prostrating, and whose prestige you broke with your patience and whose arrogance you swept away with your resolve. It is as if I am with you as you demolish its fortresses which it built on the foundations of fantasies and seduction and as you rub its nose in the dirt and tear down the facade of its civilization which had tempted those both near and far. It is as if I am with you as you hold in your hands the steel pickaxes of faith with which you follow the example of your Prophet on the day of the Manifest Victory when he crushed those deaf stones erected around the Ka'aba while repeating, "And say: truth has arrived and falsehood has perished, for falsehood is bound to perish." (17:81)

Here is the superstate - as its worshipers call it - beginning to ramble and roam among the nations, to beg them, ask their help and intercede with them in the hope that they will stand by its side, after it has been weakened by bounds, bled by war, worn down by the successive disasters and consecutive blows, driven by the horror of the battle, and after "there appeared to them from Allah what they had never expected."

So today, you truly are the people most deserving of the description given by Allah of those whom Allah honors with His expansive grace and pours upon them His prodigious favors when those who turn on their heels turn back in apostasy.

"Allah will produce a people whom He will love as they will love Him; humble with the believers, mighty against the unbelievers, fighting in the way of Allah, and never fearing the blame of a blamer. That is the grace of Allah, which He bestows on whom He pleases. And Allah is greatly possessing of grace, All-Knowing." (5:54)

And without a doubt, you, with the purity of your banner, straightness of your path and clarity of your goals, are a link connected to the successive caravans and battalions of faith from the time of our Prophet (peace be upon him) to the establishment of the Hour, as the truthful and trustworthy on informed us,

"A party from my Ummah will continue to be victorious on the truth, unharmed by those who betray them, until the decree of Allah comes while they are like that."

So rejoice at this Prophetic badge of honor which you have put on by the grace of Allah first and then by the grace of your reliance on your Lord, perseverance in your battle and steadfastness in your creed, unconcerned by the campaigns of slander meant to dirty the face of this pure and shining Jihad, uninterested in the stabs of the empty-hearted ones into whose souls love of this life has crept, and never expecting to be joined by the hesitant waverers who have been terrified by the aggressive posturing of your enemy and his constant threats.

Thus, my beloved brothers, it behooves you to be conscious of this great blessing which Allah has bestowed upon you and of whose people and place of descent Allah has made you and chosen your land to be its site, and to be the most eager of people to preserve, guard and defend it and pick its fruits through constant thanks, continuous remembrance, serious work and profound wisdom: "And [remember] when your Lord made it be known that if you are grateful, I will increase [my favors] to you; but if you show ingratitude, truly my punishment is terrible indeed." (14:7)

And it behooves you to beware in every way of it being stolen from you after having reached where you are today, by neglecting it or becoming divided over it, or by pretending to forget it with the passing of time, because blessings are only stolen and favors are only lost through what the hands earn.

"That is because Allah will never change a grace which He has bestowed on a people until they change what is in themselves, and that Allah is He who hears and knows." (8:53)

It is one of Allah's trusts which He ordered you to discharge to its people. So take it firmly and don't squander it or allow anyone else - whoever he might be - to lose it or diminish it, for the eyes of Muslims all over the world continue to observe you and prepare for the day of the Greatest Victory and Establishment whose foundation and first bricks you have laid with the founding of the Islamic State of Iraq, which came out of the womb of pains, wounds, suffering and sacrifice for it to be - with Allah's help - a newborn staring out in this dark, tyrannical world, not with shouting, wailing, condemning and criticizing, but with "there is no god but Allah" and "Allah is the greatest" and the resonance of the word of faith and monotheism.

How much the Ummah has been waiting for this day, to taste the sweetness of justice after having being fed lumps of oppression for so long, to enjoy the light of faith which it had been deprived of for all those years which it spent groping in dark tunnels full of misguided ides, perverted schools of thought and diverse whims, and to have its chest to be freed of the nightmare which sat on it for many long years of repression, humiliation, disgrace, and total vassalage and shameful submission to the lowest and most despicable of creatures.

So appreciate the value of the status which Allah has given you, then see what you will do next.

"Allah has promised those of you who believe and work righteous deeds, that He will surely grant them in the land inheritance [of power] as He granted it to those before them, and that He will establish in authority their religion which He has chosen for them, and that He will exchange their fear for security and peace. They worship Me [alone] and don't associate any with Me. And whoever rejects faith after that is rebellious and wicked." (24:55)

Beloved Mujahid brothers: your state - or rather, our state and every Muslim's state - is a newborn which the world has long awaited with the passion of the lover and yearning of the repressed, and it is in dire need of serious attention, trustworthy care and complete maintenance in order for it to grow up with a pure character and not become Nazarene, Jewish, or Magian nor run off the path in any direction whatsoever. And this requires from all of you grueling effort, total wakefulness and diligent work with which the state will develop naturally and sequentially in a way appropriate to its strength and keeping in mind its ability and combining resolve and justice, tenderness and power, comprehension and realism, wisdom and sincerity, conviction and courage, and enjoining of good and forbidding of evil with adherence to its etiquettes, norms, fundamentals and jurisprudence: "Those who, if We establish them in the land, establish Salat and give Zakat, and enjoin right and forbid wrong, and with Allah rests the end of [all] affairs." (22:41)

It also requires all sincere Mujahideen in Iraq to combine their efforts, close their ranks, and join forces to be a compact, cohesive rank and single hand with which the hearts of the true lovers are gladdened and the hearts of the jealous and hateful are enraged.

And from here, I call on and urge my Mujahid brothers in the Helpers of the Sunnah group, the Islamic Army, the Mujahideen Army and other Jihadi groups which continue to enjoy success and renown in fighting the Nazarene occupiers and their apostate hirelings: I invite both their leadership and members to the racing to goodness which the Quran calls for, encourages and orders, and in which lies the success of this world and the next: "And hold fast, all together, to the rope of Allah, and be not divided among yourselves And remember Allah's favor on you, for you were enemies and He joined your hearts in love, so that by His grace, you became brethren, and you were on the brink of the pit of fire, and He save you from it. Thus Allah makes His signs clear to you, that you may be guided." (3:103)

And I call on them to provide the best help possible in order to make this tremendous project, the Islamic

State of Iraq, successful by their joining in heart and soul with their brothers in it and standing at their side to support it and strengthen its pillars.

So just as unity is a legal obligation first and foremost, it is also a realistic demand and urgent need imposed by circumstances and required by the stage which the blessed Jihad in Iraq is passing through: "And obey Allah and his Messenger and fall into no disputes, lest you lose heart and your power depart. And persevere, for Allah is with those who persevere." (8:46)

Sticks refuse to break when banded together

But if they come apart, they brake one by one

So rise above the causes of differences and division, overcome all the obstacles which stand between you and this lofty goal, and give precedence to the interest of your religion, demand of your Ummah and protection of your march.

And I remind you of the statement of the legal expert of the Companions, Abdullah bin Mas'ud (with whom Allah was pleased),

"What you dislike in unit is better than what you like in division."

And here are your brothers in the Islamic State of Iraq (may Allah safeguard it) opening their hearts to you, holding out their hands, humbling themselves and continuing to address you as their beloved brothers about whom they are concerned, and knocking at every door in order to reach this noble goal of agreement and unity.

An we call on them to continue this good conduct which they have shown towards you with advice, lenience, humility and flexibility.

May Allah increase your guidance.

So hurry, beloved ones, to be with them as a fortified fortress with which the Shari'ah of Islam is established, under whose standard the people of monotheism are united, and which preserves the sacrifices of the heroes and efforts of the sincere ones, for you to receive through that the praise of the righteous reformers in this world and the highest of degrees in the next.

My Mujahid brothers in Iraq: today, you are the spearheads, leaders of the caravan and on the frontlines, and so your victory - with the permission of Allah, the Strong and Wise - will be a crucial one, so escalate your campaign, strengthen your resolve, increase your zeal and gird up your loins, and protect the trust and lift yourselves and your efforts above nonsensical and insignificant things abandon personal interests and beware of hidden desires, and know that your entire Ummah is behind you with hearts which hope for victory and await triumph, so don't disappoint or betray them, and demonstrate to them guidance, correctness and steadfastness on the path of Jihad and fighting and clarity of approach.

So continue the march you began, and harvest what you have sown, and avoid the mazes which will only exhaust you.

Beloved Mujahid brothers: your blessed Jihad, your great fight and your decisive battle isn't the Jihad of the Iraqi people alone, nor the Jihad of just a group or party: no, it is the Jihad of the entire Islamic Ummah. It's the Jihad of the truthful martyrs who presented for their religion, creed, and Shari'ah the most precious thing they possessed by giving heir lives - and giving up oneself is the ultimate in generosity - and presenting their skulls with generous hearts and satisfied souls, as the rows of martyrdom-seekers came one after another from every corner of the earth, racing to immerse themselves in the sea of death for the honor and happiness of their Ummah.

So Iraq deserves to be proud of them and be awarded the title of "the land of martyrdom-seeking."

It is the Jihad of the confined captive who has swallowed lump after lump of humiliation, insult, suppression and severity at the hands of the worshippers of the Cross and their hirelings, in Abu Ghrayb, Guantanamo, Bagram, Jordan, the Arabian Peninsula and elsewhere, all the while patient, seeking his reward, and awaiting the day of your victory and hour of your success to receive the reward of his sacrifice in safety and jubilation, not in worry and fear.

It is the Jihad of the homeless and scattered who have been burned and their villages destroyed by the bombs of Crusader hatred in Fallujah, Tal-Afar, Qaim and elsewhere, and are awaiting the day on which the earth is cleansed of the infidels' filth and their might is broken so that they can return to their homes and dwellings in security and confidence for the tragedies to turn into celebrations and their sadness to turn into happiness.

it is the Jihad of the pure, chaste and sheltered women whose honor was attacked by the filthiest of creatures, for the sole reason that their religion is Islam and their creed is monotheism, and who place on your shoulders a trust which weighs heavily upon the unshakeable mountains: to take revenge for them and console them in their ordeal.

It is the Jihad of the displaced fugitives of East and West for whom the earth has become narrow despite its spaciousness, and who find no refuge or shelter anywhere, as the hands of the infidels snatch them wherever they go. They look at you and their hearts long for the day of shelter and establishment on which the oppression, pursuit and displacement which almost never leaves them for so much as a second will be removed form them.

It is the Jihad of the scholars who speak the truth, those who have exposed their honor to defamation and slander for your sake. They are defending you with their pens and tongues day and night, dispelling the misconceptions of the rumormongers and refuting the allegations of the liars, and bearing for that all manner of pain and hardship.

It is the Jihad of the entire Muslim Ummah, which has been dominated by traitorous, evildoing scum and ruled by the lowliest of people from Senegal in the west to Indonesia in the east and burned by the fire of tyranny, oppression, repression, suppression of breath and imported and local laws of unbelief. They are searching for a wind of guidance with which to reinvigorate their miserable lives and restore their lost dignity.

It is the Jihad of your brothers in the Holy Land which Allah has blessed, Palestine, where their trials, severity and suffering have reached an indescribable state, and where the grandsons of monkeys and swine continue to make them taste all types of torment, including displacement, eviction, detention and demolition that turns the hair of boys white. They have placed their hope in (after Allah the Most High) your blessed Jihad, for you to be their helpers who strengthen their hand and bolster them.

So fear Allah in all of that.

Beloved Mujahid brothers: today, you are at a crossroads, as the signs of breakdown and indications of defeat of your occupying enemy have appeared on the military front, especially after his failed security plan, and his defeat - with Allah's permission - is close at hand. He has begun to wrestle with death and deal with its throes in search of a way out with which to save himself, even if it be like the eye of he needle, and he knows with certainty that he is destined to lose this battle. And the announcement of the old hag Britain, his biggest ally, about the withdrawal of some of her troops is only the beginning of the collapse and breakup which cannot be hidden by media blackout and politicians' lies.

Here is the declining enemy resorting to deceit, evasion, plotting, falsehood and slander, and running towards his lackeys in the region to extract him form the pickle on which he is drowning, and here are the threads of conspiracy beginning to be sown by their filthy hands and plotted by their devious minds.

the government of Al Saud has enter the arena of conflict, not with its emaciated army - which is too weak, flimsy and lowly to wage such battles - but by donning the cloak of concern for the Sunnis of Iraq and eagerness to extinguish the fire of Rejectionist hatred which is roasting and devouring them.

This underhanded conspiracy is designed to trick anyone for whom Jihad is not a well-founded religion, stable doctrine, established act of worship and eternal methodology, for him to fall into the trap set for him and the plot to surround him from all sides. If not, then what's this Islamic zeal and doctrinal awareness that has moved the Al Saud government and its followers, the governments of the region, to defend the Sunnis in Iraq, even though we all know that the prison of these states are choked with thousands of scholars, worshipers, students of knowledge, and Mujahideen, those with living jealously, honest help and real loyalty towards their brothers from the people of the Sunnah.

If theses emaciated hireling governments are really serious about fighting Rejectionism and its people and stamping out their project in the region, why then are they combating the criminal Rejectionists in Iraq and opening the way for every speaker to explain the truth about them, expose their plans and issue fatwas against them even as the same governments open the doors for the Rejectionists on the

Arabian Peninsula and in the city of the Messenger of Allah (peace be upon him), where their polytheistic festivals and heretical holidays are celebrated under the protection and patronage of the agencies of the same Al Saud government?!

Moreover, the Rejectionists hold the highest of position in those countries, and not just that, but anyone from whom is sensed the least opposition or anger towards the Rejectionist tide in the Arabian Peninsula is arrested and tortured. And if you wish, ask any one of their prisons, and it will tell you the real story.

"Are your believers better than those or have an immunity [from punish-ment] in the sacred books?" (54:43)

Those defending the people of the Sunnah in Iraq are the sincere Emigrants and Helpers who haven't made their Jihad a lowly commodity to be bought and sold in the markets of international interests nor subjected it to political deals which legalize it one year and prohibit it the next. They have ransomed the people of the Sunnah with their lives, defended them with all they have, and made the evildoing Rejectionists taste misfortune and swallow cups of sorrow and have torn them in every way, to give victory to the truth, not to claims of interests, and in accordance with the duty of loyalty, not the deceptions of misguidance and whims. These truthful Mujahideen will never allow anyone to make the slogan "Defense of the People of the Sunnah" an inroad through which the corrupters slip in to divert the Jihad from its course, or steal its fruits, or blur its cause, or make it a ball which the tyrants play with according to their whims and desires. This is why our Sunni brothers in beloved Iraq have no need for the filthy hands of the government of Al Saud to be extended to them; those hands which have never entered any one of the Muslims' causes without polluting, corrupting, and perverting it and misplacing its fruits, as they have done in all contemporary fields of Jihad.

Rather, our brothers are in need of the backing and aid of the Muslim peoples, with their bodies and wealth, with shelter and prayer, and with incitement.

This is the help they need, and it is the help which brings pure fruits and preserves lives, defends honor and paves the way for the purified Shari'ah, not the "Shari'ah" of Al Saud. These peoples are connected with the Sunnis in Iraq through faith-based loyalty and are trusted by them due to their innocence of the lying of politics, deception of interests and greasing of the media.

Mujahid brothers in Iraq: be fully aware of what is being plotted against your Jihad and state. It isn't like these evildoers in whose territorial waters steamboats of destruction drop anchor and on whose land lie fortified military bases from which aircraft carrying death and devastation take off, and who, with unmatched generosity, spend billions of their wealth to support your enemies, it isn't like them to offer any advice, benefit or good thing to you and your peoples, however much they pretend to, without adding enough poison to exterminate you and sabotage your Jihad, which you would only discover when it's too late and regret is of no use.

"O you who believe! If you obey the unbelievers, they will drive you back in your heels, and you will turn back losers. Nay, Allah is your protector, and He is the best of helpers." (5:149-150)

So don't be drawn in by the flashy advertisements for Satan and his allies to trap you in their twisted, tangled and knotted nets, for I swear by Allah, to do so would be nothing but a flagrant betrayal of the blood of the martyrs, a gratuitous killing of the efforts of the loyal and honest ones and a useless, pointless waste of time, and worse than that, would smooth the way for the trespassing usurper and his aides to establish himself once more and again control the necks of the people and thus start all over again.

So the matter is extremely dangerous and isn't the simple thing which those who promote it, beautify it and advertise it make it out to be. There is no way to establishment and preservation of states other than Jihad in the Path of Allah and Jihad alone, and no way to expel the trespassing, usurping occupier and crush his underlings other than taking up arms, using forces and fighting while seeking help in Allah, relying on Him and constantly asking for his aid.

This is the path, and anything else if form the whispers of Satan. Through it, the Shari'ah is established, the religion is preserved, the plots are foiled, the intrigues are revealed, the outsiders are exposed and the traitors are stamped out.

So take this path and keep tightly to it, and Allah will be with you and won't let your work go to waste.

And our final prayer is that all praise
is due to Allah, Lord of the worlds.

To The Army of Difficulty in Somalia

Release Date: 25 Mar. 2007
Production Date: Feb. 2007
Type: Video statement
Version: Transcript

To the Army of Difficulty in Somalia

Shaykh Abu Yahya al-Libi

As-Sahab Media

Muharram 1428 / February 2007

In the name of Allah, the Most Compassionate, the Most Merciful.

All praise is due to Allah, and may prayers and peace be upon the Messenger of Allah and upon his family, companions and allies.

To my brothers the Mujahideen and Murabiteen, the army of difficulty in Somalia the beloved: peace be upon you and the mercy of Allah and His blessings.

I will not be devoting my talk to bewailing a bygone past, or regretting a lost possession, even if it was gold, or grieving over something missing which cannot be brought back by grief and sorrow: "Éso that you not grieve over what escaped you nor over that which befell you. And Allah is well aware of all that you do." (3:153)

Don't grieve over what has been lost Here you have grieved, but to what benefit?

Crying and sobbing has no place in a battle in which the infidels have combined all their powers, recruited all their troops, and called out to each other, advising themselves thus: "And their chiefs went out quickly [saying], 'Go on, and remain devoted to your gods, for this is certainly a thing plotted [against you].'" (38:6)

And they became enraged as they watched a seedling of life begin to flourish, come alive and shine in that barren, barefaced desert, and it was then that "they said, 'Swear a mutual oath by Allah that we shall attack him and his people at night, and that we shall then say to his heir: we were not present at the slaughter of his people, and we are telling the truth.'" (27:49) And they came out with their massive armies "recklessly and to be seen by the people, and to hinder from the path of Allah, and Allah encompasses all that they do." (8:47) Their hearts were full of arrogance, their chests overflowed with fury, their noses were turned up in pride, and their minds were lost in satanic whispers: "And when Satan beautified to them their [evil] deeds, and said, 'No one among men can overcome you today, while I am with you.'" (8:48) Allah permitting, they will soon discover - as their predecessors and allies have discovered - that their plots are passing fantasies and their forces are empty, powerless assemblages: "But when the two forces came in sight of each other, he turned on his heels [in flight], and said, "I am innocent of you; I see what you see not; [and] I fear Allah, for Allah is strict in punishment.'" (8:48)

It is not for the people of tyranny and oppression to be resolute against the champions of doctrine, faith and conviction, and their plotting shall backfire and be a curse for them: "But the plotting of evil will afflict only its [plotters]. Are they waiting for what happened to the first ones? And no change will you find in Allah's way [of dealing], and no diversion will you find in Allah's way." (35:43)

And their tyranny will be the source of their doom: "That, and whoever punishes as he was punished, and then is [again] oppressed, Allah will help him, and Allah is Most Pardoning, Most Forgiving." (22:60)

Allah has decreed that oppression fells its people And that the tables are turned on the tyrant

One verse of Allah's Noble Book is sufficient to console us over our tragedy, bandage our wounds, whet our appetites and make our feet firm: "And lose not heart, nor fall into despair, for you must gain mastery if you are [truly] believers. If a wound has touched you, be sure a similar wound has touched the [other] people. And those are the days [of varying fortunes] We give to people by turns, that Allah may know those that believe, and that He may take for Himself from your ranks witnesses [martyrs]. And Allah does not like those that do wrong." (3:139-140)

And another verse encourages us to continue on our path, pours into our hearts the concept of fostering perseverance, motivates us to make ourselves - despite the wounds and disasters - the pursuers instead of the pursued, and shakes from us the dust of weakness, frailty and impotence: "And do not lose heart in seeking the enemies. If you are hurting, they are hurting as you hurt: but you hope from Allah what they hope not. And Allah is all-Knowing, Wise." (4:104)

So O lions of Somalia and champions of the deserts and jungles, O groups of the army of difficulty: resolution and perseverance, for by Allah, this is merely a distress which will vanish and a darkness which will pass, to be followed by nothing less than certain victory, pure consolidation and good outcomes. I tell you what the Prophet of Allah Moses (peace and prayers be upon him and our Prophet) told his people after they had been shaken by trials and Pharoah and his legions had inflicted on them all manner of punishment and torture and had slaughtered their sons and left their women alive: "Said Moses to his people, 'Pray for help from Allah, and be patient, for the earth is Allah's, to give as a heritage to such of his servants as He pleases; and the end is for the righteous.'" (7:128) And what was the outcome of their patience after the severity of their trial? "And We made a people who were oppressed inheritors of the East and West of the land upon which We sent down Our blessings. And the good [promise] of your Lord was fulfilled for the Children of Israel because they had patience, and We destroyed what Pharaoh and his people had built." (7:137) And the favor of Allah came to them from where they least expected it and His mercy enveloped them and He aided them with His victory after they said, "We will be overtaken" (26:61), and they became the inheritors and He destroyed their enemy in a blink of an eye. "And We wished to bestow our favor on those who were being oppressed in the land and make them leaders and make them inheritors, and establish them in the land, and show Pharaoh, Haman and their troops what they had been fearing from them." (28:5-6)

So after severity, there is only comfort, and after difficulty, there is nothing but ease, and after hardship, there is nothing but relief: "So verily, with every difficulty, there is relief. Verily, with every difficulty there is relief." (94:5-6)

If difficulty becomes severe, then hope for ease, for Allah's decree is that difficulty is followed by ease Allah will bring relief, for He Orders His creation every day What you see will not last, and you will see Relief from what time insisted on So when time brings difficulty, be firm And patient, for the key to firmness is patience How many a long-held worry was lifted And another's difficult straits easedÊ

There are only two parties, and they have no third: the party of truth which fights under its banner and for it, and the party of unbelief which aids its darkness and spreads its misguidance. And what a difference there is between the mirage and the water. "Those who believe fight in the way of Allah, and those who reject faith fight in the way of the false god. So fight against the friends of Satan; feeble indeed is the plotting of Satan." (4:76)

So rejoice, O guardians of Tawheed, and be hopeful, and shed the feelings of despair, blow away the clouds of despondency and repel the illnesses of impotence. And know that Allah is with you, surrounding you with His protection, defending you with His strength, and helping you with His help. As for the Abyssinian rabble, it suffices them that their ally is Satan. And their eyes have been blinded, so they haven't learned from what their brothers - nay, masters - are suffering in Afghanistan, Iraq, Chechnya, Algeria and elsewhere. "Do they not travel through the earth, and see how the end of those before them was? Allah brought utter destruction on them, and the same [will be] for the unbelievers. That is because Allah is the Protector of those who believe, and because the unbelievers have no protector." (47:10-11)

Isn't Allah, the King of Kings and Lord of Lords in whose hand is supreme power over everything and who protects and from whom there is no protection, the One who is defending you? What, therefore, can these lowly ones do to you? "Verily, Allah defends those who believe. Verily, Allah does not love every unfaithful, ungrateful one." (22:38)

And accept the good news of your Prophet (peace be upon him), and be certain of its truth. Said he, "Give this Ummah the good news of eminence, religion, superiority, victory and establishment in the land."

So reaching this noble goal - i.e., establishment in the land, victory over the occupiers, and superiority and eminence in religion - is a certain thing, but for who? For whoever takes the straight path, relies on the Strong and Merciful and grasps the surest handhold, however long the march and difficult the situation: "O you who believe! If you help Allah, He will help you, and plant your feet firmly." (47:7)

And in order for you to guarantee a good outcome - Allah permitting - and protect the route of the caravan of truth and safeguard the Muslim group from wasting its efforts and going astray, you must grasp the devotion of Jihad and hold tightly to it. And I bring attention to the word "devotion," for by Allah, Jihad is not a "choice," as some have degraded it by adding this ugly word to it and saying (and what a terrible thing they say) "the choice of Jihad" and "the choice of resistance," thus dirtying its face and fiddling with its meaning. Jihad is a prescribed, obligatory devotion made compulsory by the Lord of Lords, He Who sent down the Book from above the seven heavens: "Fighting is prescribed for you, and you dislike it." (2:216)

We will not be like those who cut up the Quran, lest we become like those who "distort the [rules of the] words after their placing. They say, 'If you are given this, take it, but if not, beware!'" (5:41) nor like those about whom Allah said, "Have you not seen those who were told, 'Hold back your hands [from fighting] and establish regular prayers and pay regular alms,' [but] when [at length] the order for fighting was issued to them, a group of them feared men as much as - or even more than - they feared Allah. They said, 'Our Lord! Why have you ordered us to fight? Would that you had granted us respite for a little while.' Say: little is the enjoyment of this world, and the hereafter is the best for those who do right; and you will not be wronged in the very least." (4:77) There is no choice of prayer, no choice of Hajj, no choice of fasting, and also no choice of Jihad. There is only "we have heard and obeyed" and "we believe in it, the whole of it is from our Lord." And let your slogan be the slogan of the Emigrants and Helpers, who understood that there is no life for the believer, no standing for his religion, and no establishment for his Lord's Law without Jihad, so declare it as a pledge which is not retracted and which is not said just anytime:

We are those who pledged to Muhammad To perform Jihad for as long as we remain

So make every effort in carrying out this great act of worship out of devotion to your Lord and perseverance in difficulty, and to preserve unity between you, and continue on your pristine path, be resolute against the advance of your enemies, and be harsh with the apostates and their helpers. "O you who believe! Fight the unbelievers who are near to you and let them find harshness in you; and know that Allah is with those who fear Him." (9:123)

The Abyssinian rabble didn't enter your land and occupy your country through conferences, negotiations, accords or talks. No, they declared a blatant war against you, for which they readied armies and came together to wage it and sought help in it from partners. So don't let your enemies be better versed in Allah's ways than you, for they have understood that armed truth is only confronted by armed falsehood. And it is not for them - after sacrificing to get to where they are today the heads of their soldiers and treasures of their storehouses - to abandon all that willingly and in submission and hand leadership over to you through mere dialogue, meetings or negotiations. So it's blood and destruction, and know that existence in this world is for the strong, so stick to strength, combat and slaughter of the people of unbelief and misguidance. And fight a guerrilla war, for it is the most durable of battles and the least in losses, most crushing, and most appropriate for the weak and few. Carry out against them raid after raid, lay ambushes for them, shake the earth beneath them with bombs and mines, destroy their bases and fortresses with martyrdom operations and car bombs, cut their supply routes, frighten with them those behind them, disturb their security, and make their life miserable, for the blood of one of them is the blood of a dog.

Light a fire and make a volcano erupt under the feet of the invading occupiers regardless of their creed and their cover, and whether they invaded your country on the back of tanks and with the power of iron and fire, as Nazarene Ethiopia did, or came to you under the cloak of international legitimacy, Security Council resolutions, peacekeeping forces, or the African Union. There is only unbelief and a codified, coated occupation with which they kill jealousy and protect themselves from the intifada, so don't be deceived by their slogans and don't let Satan lure you into their traps. And seek protection in Allah: He is your Protector, and the best of protectors and best of helpers.

My brother Mujahideen in Somalia: beware of those who climb over the scattered limbs of the people of sacrifice. Beware of the bandits and robbers of efforts, those who satisfy themselves with what they were not given; the sharp-tongued impeders who sit on the fence, and whenever the balance swings in someone's favor, they run towards its people and swear to them that they are with them; "Wavering between [this and] that, [belonging] neither to these nor those." (4:143) "Those who await your [fate], so if you gain a victory from Allah, they say, 'Weren't we with you?" whereas if the unbelievers gain a victory, they say [to them], 'Didn't we surround you and guard you from the believers?'" (4:141)

Those who, every time their condition is exposed and their affair becomes clear, make excuses, and their excuse is always, "They swear by Allah, 'We only wanted kindness and reconciliation!" (4:62) They are the most dangerous group for the Jihad and Mujahideen, and they are their affliction in all arenas of modern Jihad. How many a banner they have toppled, and how many a fruit they have stolen, and how many a methodology they have perverted, and how many a Jihad they have wasted. They are the discouragers of resolves, slow to go forth, quick on the day of booty, those who blur the issues, those who are strangers to Jihad and not from its people, and by Allah, the likes of these, however much they camouflage themselves with equanimity and pretend to be concerned about the greater interest, "If they had gone out with you, they would only have provided you with corruption, and would gone back and forth between you [with evil talk] seeking sedition among you, and there are among you those who would have listened to them. And Allah knows the wrongdoers." (9:47)

So take the advice of your Lord regarding them and their like: "So if Allah brings you back to a party of them, and they ask your permission to go out, say: you shall never go out with me, nor fight an enemy with me. You preferred to sit the first time, so [now] sit with those who stay behind." (9:83).

And their identifying mark is fickleness and inconsistency. Thus, sometimes they come in the dress of wisdom, and at other times they wear the clothes of rationality. At times they ride on the back of political sophistication, while at other times - and this is most often the case - they feign moderation, centrism and balance, although it's nothing but cowardice, feebleness and love of this world.

The cowards consider impotence to be smart And that is the deception of lowly nature

So beware of them.

My patient brother Mujahideen in Somalia: protect the clarity of your banner and the purity of your path, and state at every turn the truth about your objectives, for the peace of mind of you and your supporters. And declare with a frankness that resonates that the goal of your fight and the purpose of your Jihad is the expulsion of the occupier and his helpers and the establishment of an Islamic state in the land of Somalia which rules by Shari'ah and rejects false gods, and only obeys the Lord of Lords. These are two inseparable things: i.e., expelling the occupier and establishing the state of Islam. And it is through this that Allah's help comes down, His aid reaches you, the hearts of your supporters are reassured, and the path is severed in front of those who climb over your sacrifices.

Champions of the army of difficulty in Somalia: you must depend completely on Allah (the Glorious and Exalted), for He is the one who has power over everything, and honor is for Him, His Messenger and the believers. And beware of expecting help from the East or West, or rejoicing in the aid of this state or that, or being deceived by the lies of the leaders of unbelief who hide in themselves what they don't reveal to you. And beware of overconfidence in those who were led by their interests to claim to stand at your side and adopt your cause, for by Allah, the hatred and animosity they carry towards you is almost as much of that of the Abyssinian trespassers, and as soon as what motivated them to back you disappears or is replaced, they will turn hostile to you, bare their fangs and reveal their true selves. Then you will go and look for others, and so on and so forth, and in this way efforts will be wasted, unity will be torn apart, and animosities will spread, and your condition will be like he who seeks refuge from heat in fire. "The parable of those who take protectors other that Allah is that of the spider, who takes for itself a house; and truly, the flimsiest of houses is the spider's house, if they but knew." (29:41)

So don't take from among them intimates from whom you seek advise in your affairs, reveal your secrets, and tie them to your cause, for them to whisper to you as they like and lead you wherever they want and dictate to you whatever they wish, for you to follow them in all that with interpretations and claims of balancing and stabilizing, and so waste your Jihad and the fruits of your efforts in these tunnels. "O you who believe! Take not others as intimates: they will not fail to corrupt you. They wish for you severe harm; hatred and animosity has already appeared from their mouths, and what their hearts conceal is worse. We have made plain to you the signs, if you comprehend." (3:118)

Perseverant Murabit brothers in Somalia: you must observe patience and encourage it, and don't be enfeebled by the great number of sacrifices, nor the continuous disasters, nor the fewness or absence of helpers, nor the length of the path, because victory comes with patience. And increase your prayers and beseeching of your Lord, and ask the weak and impoverished to do the same, for that is one of the greatest causes of victory, as the Chosen One (peace be upon him) said, "Allah gives victory to this Ummah through its weak ones: through their supplications, prayers and devotion."

I also ask my Muslim brothers to stand with their brothers and go forth to fight at their side, because all the things which make Jihad an individual duty are present in their battle against the Abyssinian occupiers and their apostate lackeys: meeting of ranks, clashing of armies and mobilization by the commanders who have appealed to the Muslims for help, aroused the fighters and sought the assistance of the Mujahideen after their land was invaded by a devious infidel enemy who took it over and ruled its people. So go forth and go forth quickly, lightly or heavily, in groups or individually, with self, wealth, prayers, garments and encouragement: "O you who believe! What is the matter with you, that, when you are asked to go forth in the cause of Allah, you cling heavily to the earth? Do you prefer the life of this world to the hereafter? But little is the comfort of this life as compared with the hereafter. Unless you go forth, He will punish you with a painful punishment, and put others in your place. But Him you would not harm in the least, for Allah has power over all things." (9:38-39)

So for whoever Afghanistan was too far away, or had the way to Iraq closed for him, or the doors to Algeria locked in front of him, or was unable to reach the land of Chechnya or had the path to Palestine shut in his face, here is Somalia, just beginning with an enemy occupier, so let him show Allah what He loves to see in him, and let him demonstrate the truth of his loyalty to Allah, His Messenger and the believers, and hurry with the lightness of a bird without making excuses or procrastinating.

And I make a special request of the Mujahideen on the information frontline, who are carrying out an act of worship which is among the most honorable of devotions and loftiest of acts of obedience: incitement to fighting, which the Lord of Honor entrusted to His Prophet (peace be upon him) and commanded him to perform, due to its great rank and importance. He told him, "O Prophet! Rouse the believers to the fight." (8:65) And He also told him, "So fight in Allah's Cause - you are held responsible only for yourself - and rouse the believers. It may be that Allah will restrain the might of the unbelievers. And Allah is strongest in might and strongest in punishment." (4:84) It is a request to them to adopt the cause of their brothers the Mujahideen in Somalia (or rather, their own cause, because the believer is to the believer like a building, one part of which supports the other), by spreading their news, broadcasting their sacrifices, following their developments, and bringing them to the Muslims through every legitimate means, whether text, sound or video, and encouraging them to stand at their side and provide them with every backing in order for the cause of Jihad in Somalia to remain alive and active, lest it be forgotten with time and neglected like other modern Islamic causes have been. How many a concern your releases have awakened, and how much hope they have revived, and how much resolution has been renewed by them, and how many men they have mobilized, and how many an enemy they have disappointed. May Allah bless you lions of the front, for by Allah, the fruits of your combined efforts - sound, video and text - are more severe for the infidels and their lackeys than the falling of rockets and missiles on their heads. So make your intentions sincere, purify your hearts, excel in your work, continue the march, and don't tire or become bored, lest Islam be attacked from your flank (may Allah protect you from that).

Anas (Allah was pleased with him) related that the Prophet (peace be upon him) said, "Perform Jihad against the polytheists with your wealth, selves and tongues." And praise is for Allah, first and last.

Palestine, an Alarming Scream and a Warning Cry

Release Date: 29 Apr. 2007
Production Date: Apr./May 2007
Type: Video statement
Version: Translation

I seek refuge in God from the evils of the accursed Satan.

In the name of God, the Merciful, the Compassionate.

"The believers, men and women, are protectors one of another: they enjoin what is just, and forbid what is evil: they observe regular prayers, practice regular charity, and obey Allah and His Messenger. On them will Allah pour His mercy: for Allah is Exalted in power, Wise" [Koranic verse; al-Tawbah 9:71].

In the name of God, the Merciful, the Compassionate.

Praise be to God. Prayers and peace be upon our Prophet Mohammed and upon his family, companions, and followers.

Oh Ummah of Islam, Ummah of sacrifice and courage, Ummah of steadfastness and loyalty, peace and God's mercy and blessings be upon you.

God, the Great and Almighty, whose words are the most truthful and the best, said: "The Believers, men and women, are protectors one of another: they enjoin what is just, and forbid what is evil: they observe regular prayers, practice regular charity, and obey Allah and His Messenger. On them will Allah pour His mercy: for Allah is Exalted in power, Wise" [Koranic verse; al-Tawbah 9:71].

Based on the principle of mutual advice in religion, which is one of its firmly established pillars, and in response to the call of the faith-based loyalty, which is represented by enjoining right and forbidding wrong and from which emanate advising the truth and cooperating to implement it, these are words I send from my loving heart to my brothers in the holy and blessed land, the stolen Palestine, whose blood continues to flow as a result of the attacks of the occupying Jews until it was surprised by a sharp dagger penetrating through to its vulnerable points and settling inside it, afflicting it with a disaster which doubled its ordeal and weakened its power. That occurred following the mistake made by the HAMAS politicians in which the feet stumbled and they led their group into an errant trap, from which they could not exit or escape from the deception of its mirages, except with the awakening from the dreams of deep sleep with which they anesthetized themselves and disappointed their followers and supporters and by the clear and evident return to the methodology of jihad in the battlefields, from which they crept away bit by bit to seek refuge until they arrived at this bottomless abyss under the cover of the political process and the call for the maintenance of national unity.

Prior to that, I would like to say that we are Muslims. We are united by Islam's true creed. We are bound together by the strong bond of loyalty. We are strengthened by the bond of faith-based brotherhood. This brotherhood requires -- when we are sincere about it and focused only upon it -- providing counsel and also accepting it.

God, the Great and Almighty said: "Let there arise out of you a band of people inviting to all that is good, enjoining what is right, and forbidding what is wrong: They are the ones to attain felicity" [Koranic verse; al-Imran 3:104]. Jarir Ibn-Abdallah al-Bajali, may God be pleased with him, said: "I gave the pledge of allegiance to Allah's apostle to offer prayers, pay the zakat [obligatory alms], and be sincere and true to every Muslim." Therefore, a Muslim may not, no matter how much he advances, look down upon listening to advice if he is serious about seeking the truth and careful to follow it. [He may not] create between himself and the truth veils and obstacles with which to disguise and take as excuses that free him from accepting the truth that reaches him, from whatever direction it comes, as long as what is being said to him and the advice he receives are honest calls, true guidance, and evident truths.

The wise word is the aim of the believer, and he is most deserving of it when he finds it. However, if a person, with the eloquence of his tongue and the strength of his declaration is capable of convincing the masses of his misguidance, obscurities, and excuses, which he above all knows are untrue, then what will he be able to tell his God, the Almighty and Most-High, tomorrow, when the secrets are exposed for judgment and when he does not have any power or support? Does he not fear, having confused the masses about their faith and embellished falsehood for them until they accepted it and followed it? Does he not fear that his fate might be among those who God described as: "Let them bear, on the Day of Judgment, their own burdens in full, and also (something) of the burdens of those without knowledge, whom they misled. Alas, how grievous the burdens they will bear" [Koranic verse; al-Nahl 16:25].

I said these words because we have constantly heard from the HAMAS political leadership whenever our leadership provided advice or instructions or warnings of pitfalls and deviations that "our methodology differs from the methodology of the al-Qaeda Organization," or "we do not need the advice of the leadership of al-Qaeda." These sentences, which have no place in Sharia, are obstacles with which to repel truth and prevent the seeking of guidance from it. Among peculiarities of the divergences in this matter is the fact that these types of statements were uttered by some of their leaders while they are in the heart of the capital of corruption and non-belief, Moscow, and within a period during which they were praising the results of their meeting with killers among the non-believing Russian leadership, whose criminality and corruption are no less than the crimes of Sharon and his party. Indeed, they described these meetings as positive and fruitful. Oh politicians of HAMAS, is the tyrant Putin and his thoughtless party, according to you and by your definition, more concerned about the interests of Palestine and the Palestinians than the mujahid Sheikh Ayman al-Zawahiri, may God protect him? And what sort of a bond, oh leaders of the Islamic resistance movement, and I say the Islamic [resistance movement], that binds you with the leaders of atheism and destruction in Moscow so as to listen to their counsel, to believe what they say, to accept their stands, and to agree with their suggestions. Is it Putin's love and that of his state for Palestine, all of Palestine, and his quest to oust the rotten Zionist occupier from its land? Or is it his serious desire to achieve security and stability for the unfortunate people of Palestine? Or is it his strong enmity for the Zionist state and his love and support of you?

Didn't you learn from those organizations and states that followed the east once and the west next, until they became very lost, not knowing which one to go after, and likewise those who accepted support from anyone but from God and spent their lives in lies, evasion, and false hope? Such states and organizations were only able to bring to their nations more humiliation, and they forced their nations to become tails in the service of their enemies and stripped them of the ability to feel proud of their identity. Do you want to take your nation back and repeat what happened in the past after it just began to regain its dignity? Praise be to God. There are those who take the infidels as allies instead of believers and who seek glory from them, but those who believe in God, His messenger, and the believers are the victorious. Do you read in the book of God, "O ye who believe! Take not into your intimacy those outside your ranks: They will not fail to corrupt you. They only desire your ruin: Rank hatred has already appeared from their mouths: What their hearts conceal is far worse. We have made plain to you the Signs, if ye have wisdom" [Koranic verse, 3:118].

So, here you are, oh HAMAS politicians, you were so disparate that you nominated Christian nobodies to be within your Muslim groups, a move that damaged your movement's labor, wasted your youth's sacrifices, entered into this dangerous situation, and placed your trust and confidence with and made the source of your advice the imams of infidelity and the heads of tyranny and atheism.

Where are you going? Did this atheist who destroyed a Muslim country and annihilated a whole people in Chechnya and dispersed its people into the valleys and the deserts in a way that surpassed the Jews' dispersion of our Muslim people in Palestine become a friend of yours? Did he become closer to you, oh HAMAS leaders, than the people who have proven their loyalty to God, His messenger, and the believers and demonstrated both in word and deed their adoption of the Palestinian issue and directed their efforts to ease the pain of its Muslim people, not as a favor to them, but as an expression of their sense of religious duty, the source of all their decisions, policies, and stands, which also should be yours?

I know that friendly, caring, and passionate people will say, didn't you find anything but the HAMAS movement which stands today with its youth and heroes against the Zionist program to address and to offer advice, precisely at this sad and difficult juncture when forces inside and outside are cornering them and when HAMAS is facing a deadly economic and political embargo? To those I say:

First, the function of sincere advice doesn't exclude anyone, and everyone and every group can use it. The prophet, prayers and peace be upon him, has said, "Religion is advice, religion is advice, religion is advice; you give advice for the sake of God, His book, His messenger, the Muslim imams, and for the sake of Muslims in general." Without advice, religion would not have survived and would have turned into a distorted Sharia by the innovators, sullied by the whim of the people of personal opinions, and torn apart by the output of philosophers and sages, and it would have stalled by the rulers and the tyrants. Religion would not have achieved its ease, clarity, and purity without the advice of the advisors and without the assessment of the righteous people.

Just as the holy land does not consecrate anyone--man's deeds do--our master Salman al-Farisi, may God be pleased with him, said, "So as the names do not protect or honor anyone, the man's deeds venerate him." "Verily the most honoured of you in the sight of Allah is (he who is) the most righteous of you. And Allah has full knowledge and is well acquainted (with all things)"[Koranic verse, 49:13].

And we say yes, HAMAS is in trouble, but who pushed it into trouble and placed it between the grind of the public's demands, the whip of international pressure, and the temptations of the Arab concession until the optimum principle and steadfastness became its rejection of the recognition of Israel rather than Israel's destruction altogether and the return to Muslim rule of every speck of land? Therefore, the purpose behind what we have said is to extricate the movement from the predicament into which it has plunged itself. Our demand is that it should rebuff this bad course which it entered without any Sharia or logical justification.

I say, second, in spite of the fact that we were watching the many dangerous pitfalls that HAMAS leaders were falling into from time to time through their policies, declarations, meetings, and stands, and in spite of our knowing that the course which they adopted would only waste the efforts and increase further the deviations and the compromises, we still wanted to bring the issue to reality and clarity so as to expose to the fair-minded the deviations of the path they are taking and to make clear and obvious their abandonment of the slogans which they used to chant and call for to motivate the people of Palestine.

When it became unbearable and their actions began to take the Muslim nation back to square one and their main goal became limited to strengthening the new secularism in the minds and the reality of the people, the truth became clear to anyone who is seeking the obvious and indisputable truth. It became our duty and the duty of every Muslim to salvage the sacrifices of this free people from the games and the tricks played by those who are pursuing high posts and those who are in charge of titles. Our silence will mean that we are partners in this evil action, which wastes the religion, alters the truth, supports the humiliating evil, and robs the honest and the sincere of their efforts. We seek refuge in Allah from betraying the blood of the heroes and the martyrs. The Almighty, may He be exalted, said, "O ye that believe! Betray not the trust of Allah and the Messenger, nor misappropriate knowingly things entrusted to you" [Koranic verse; al-Anfal 8:27].

What we have to do is enlighten the people with the truth and cast the fallacies with it. Our purpose is to please God, even if it means displeasing the entire world. For those who proclaim the messages of God and fear Him, do not fear anyone but Him, and God is sufficient unto us.

The prophet, may prayers and peace of God be upon him, said, "Whoever pleases God by displeasing people, God is sufficient unto him. But whoever displeases God to please the people, God left him to the people to deal with him." The name of your Islamic resistance movement requires that the movement holds on to the path of jihad, which you had called resistance, and in spite of our reservations about the name, it can only mean that. Besides, the founders of this movement wanted it to be an Islamic movement and there is no meaning to its Islamism unless it differs completely from the rest of the nationalistic and patriotic movements by total commitment to the rulings of Islam, inside out, in its knowledge, its creed, its Da'wah, its inciting others, its jihad, its policies, its relations, and its statements. "Say: Truly, my prayer and my service of sacrifice, my life and my death, are (all) for Allah, the Cherisher of the Worlds. No partner hath He: this am I commanded, and I am the first of those who bow to His will" [Koranic verse; al-An'am

6:162-163].

By this we differentiate between the path of the criminals and the path of the believers in all the issues and their branches, their entirety, and their details. By it, the banners are distinguished and the goals are set so that they do not remain floating and unclear, where it is hard to tell right from wrong and the honest from the dishonest. In this way we explain the verses and recognize the path of the criminals. On this basis, the origin of your movement and its establishment was to make all religion for God and not to divide it and choose from it what suits the soul and reject the rest, under the pretext of interests, policies, and the requirements of the time. How can you make all religion for God the essence of your resounding speeches and dazzling slogans and move the people and awaken enthusiasm among the youth and then as soon as you knock at the doors of the twisted and lame politics, take cover under the dome of the polytheist Palestinian Legislative Council and meet with the infidels and travel east and west; you shun it [Islam] with your deeds and you insult it with your words? Tell us then; which Islam are you talking about and what Sharia is that which you intend to establish and implement as you claim?

"Out of what Allah hath produced in abundance in tilth and in cattle, they assigned Him a share: they say, according to their fancies: 'This is for Allah and this - for our partners!' But the share of their 'partners' reacheth not Allah, whilst the share of Allah reacheth their 'partners'! Evil (and unjust) is their assignment" [Koranic verse; al-An'am 6:136].

Religion has become so mutilated and so defaced and subject to preferences. It is a religion that does not exist except in the world of fallacies and propaganda, and it lives in the memory of illusions. "Such is (only) your (manner of) speech by your mouths. But Allah tells (you) the Truth, and He shows the (right) Way" [Koranic verse; al-Ahzab 33:4]. So, where is your religion, oh leaders of HAMAS, from the case of implementing the Sharia, all the Sharia which you slaughtered with your own hands, when you agreed to follow the infidel religion of democracy, which is founded on the basis of the rule and sovereignty of the people. The religion of God says, "The command is for none but Allah. He hath commanded that ye worship none but Him: that is the right religion, but most men understand not" [Koranic verse; Yusuf 12:40], but the religion of democracy, which has decayed the drained body of our ummah, states that the rule is for the people, and that no one else has a say in it but the people. According to its religion, no one can comment on the people's rule or refuse the people's judgment. Which of the two religions do you choose, and which of the two commands do you follow? Where is your religion, oh leaders of HAMAS, from God's area of jurisdiction, the religion of Whom you claim to want to establish from the right to legislate, the structure of which you have destroyed and the pillars of which you have removed in the name of politics, diplomacy, worldly wisdom, and maturity of thought, when you placed yourselves under the dome of the parliament to legislate and compete with God over this right? That evil dome, which you gave an oath to respect, has been given the right, under the secular constitution, to allow and forbid, legislate and impose. It permits what its people find permissible, even if it were forbidden in God's religion, and it forbids what its masters have forbidden, even if it were permissible in the Sharia of God. The religion of God says, "Is it not His to create and to govern? Blessed be Allah, the Cherisher and Sustainer of the worlds" [Koranic verse; al-A'raf 7:54]. It also says, "It is not fitting for a Believer, man or woman, when a matter has been decided by Allah and His Messenger to have any option about their decision: if any one disobeys Allah and His Messenger, he is indeed on a clearly wrong Path" [al-Ahzab 33:36].

Your polytheist legislative council, of which you agreed to be among its top figures, states: But, we are the gods who have a say. There will be no rule except what we decide and no law other than what we agree to and no coercion except in what we enforce and no punishment except against those who violate our legislation.

We, as members, have a choice in deciding our matters even if God and His messenger decide otherwise. We will not lie to you or to your council composed of conflicting partners. Tell us then, isn't what we said the truth and the core of the matter? Yes it is even if some claim otherwise. Can your Islamic movement legislate for the Palestinian people one Sharia law without first putting it through the council of apostasy that blocks God's law? Then what is the matter with you? How do you arrive at your judgment? You should know that names will not change the truth no matter how nice they are made to sound and will never be enough justification to follow falsehood, or to escape God's orders, or to avoid accountability in this life and the hereafter. The prophet, may prayers and peace be upon him, "From among my nation there will br people who will drink alcohol calling it by other than its real name." Misguidance is misguidance, disbelief is disbelief, legislation is legislation, and manipulation of religion is manipulation, whether they call it politics, shrewdness, interest [maslahah], consideration of national interests, or following Arab consensus. "And most of them follow nothing but conjecture. Certainly, conjecture can be of no avail against the truth. Surely, Allah is All-Aware of what they do" [Koranic verse; Yunus, 10:36]

What is your religious stance, oh leaders of HAMAS, on the issue of allegiance and enmity [wala wal bara] which is the core of Islam, its spirit, its solid pillar, its strong foundation, and the strongest bond in Islam, the breaking of which breaks all bonds. Breaking this bond undermines laws turning them into an object of play and subject to distortion by fools. Without these laws, corruption fills the world, strife spreads, and calamities continue. This is what you are subjected to today at the hand of your kinsmen who hate you but whose friendship you seek. "And those who disbelieve are allies of one another, if you do not do so, there will be Fitnah [strife] and oppression on the earth, and a great mischief and corruption" [Koranic verse; al-Anfal, 8:73].

Islam has made the bond of wala and the bond of faith the same bond. [God] said: "The believers are but a single brotherhood" [Koranic verse; al-Hujurat, 49:10]. It is therefore the axis of it all. The believers are brothers whether they are close or distant, Arab or non-Arab, black or white, man or woman. The prophet, may prayers and peace be upon him, announced this fundamental principle to the largest gathering of Muslims when he said, "Oh people, your Lord is one, your father [ancestor] is one; no Arab has any superiority over a non-Arab or a non-Arab over an Arab; no red is superior to a black, nor a black superior to a red except by their righteousness. Have I conveyed the message?" The prophet asked and the people answered, "Indeed so, oh messenger of God." The prophet then said, "Let him that is present tell it unto those absent."

Why is it then that we are today hearing you speak of other bonds and new alternatives that have nothing to do with Islam? Nationalism and Palestinian unity and other things have become the foundation of your relations and the basis of your ties. You used them to get through thick and thin and through them you entered into the territory of misguidance, falling into falsehood and tainting your group which was purified by the blood of martyrs. One who listens to your interviews and statements can longer differentiate between you and the secular movements; you declared them as your brothers and befriended and allied yourselves with their leaders and followers. Thanks to this deviant path, the Baha'i servant of the White House [Abbas] is now referred to as Mr., president, brother, and his excellency, while in an effort to please the atheists, you decided that the Chechen cause was an internal issue that had nothing to do with you as one of your politicians stated. This is the product of the jurisprudence of the despicable political dealing that misguided the minds and shook the religious principles. The jurisprudence of Islam and the guidance of the prophet we use as the guiding light said, "The believers are like one body; when one part of it suffers, all other parts join in, sharing the sleeplessness and fever." God Almighty said, "You will not find any people who believe in Allah and the Last Day, making friendship with those who oppose Allah and His Messenger even if they were their fathers, or their sons, or their brothers, or their kindred" [Koranic verse; al-Mujadalah, 58:22]. God Almighty also said, "O you who believe, take not for supporters and helpers your fathers and your brothers if they prefer disbelief to Belief. And whoever of you does so is one of the wrongdoers" [Koranic verse; Tawbah, 9:23]. The prophet, may prayers and peace be upon him, said, "The best bond in faith is alliance in God, love in God, and hate in God Almighty" [hadith].

Your embarking on this dangerous path without concern or worry is what made the rejectionists in Iran, the enemies of the companions of the prophet and the enemies of Islam, your brothers whom you flatter at the expense of your faith and the fundamental principles of your religion. You glorified their imams and idolized the graves of their filth with your visits. You opened the doors for them to spread Shiism among the Muslim Sunni Palestinian people and allowed their corrupt hands through your group to spread vice in the holy land, destroying people's faith, muddying their thinking, and distorting their nature all under the pretext of the general interest, the delusion of need, and the lies of necessity. You continue to turn a blind eye to their historic and current crimes which total, by God, no fewer than the crimes of the criminal Jews. Your case is like the case of someone seeking refuge from heat in fire. On the other hand, you turned away from your brothers in faith and methodology and from the truthful mujahideen who are honest when they give advice and sincere when they give support and just when they pass judgment. They distance themselves from deviant temptations and sayings of the lords of corruption. "(So) which of the two parties has more right to be in security? If you but know. It is those who believe and confuse not their belief with wrong, for them there is security and they are the guided" [Koranic verse; al-An'am, 6:81,82].

I am addressing those youthful and pure young men within the brave Qassam brigades; the students of the lion, Yahyah Ayyash, God rest his soul; those who are faithful to the blood of their martyred brothers and who, with rare courage, have set the most magnificent examples in the different types of sacrifice in the stories of boldness, honest fervor, and bravery, the like of which is venerated by heroes; I address them as a sympathizer and I say to you, Oh you numerous lions! Your faithfulness to the blood of your predecessors of honest men will only be accomplished by strict adherence to the path of jihad, battle, and strikes, for which your brigades were established, the sole purpose to revive it, to strengthen it, to keep it going, and to reject all paths besides it, no matter how it is embellished or alluded to by others or sought after by the defeated because the truth has appeared and the deceptions have vanished. It is like the froth of the sea, which disperses at the first sign of turmoil and gets scattered by the smallest storm, disappearing as if it never existed as froth always does, for frost dries up and that which is beneficial to man always remains. That is God's way of presenting us with examples.

Oh, Qassam Brigades! Where is the revenge? Where are the fires? Where are the bombs? Where is Ibn-Ayyash to renew your glory and show us the enemy's towers collapsing? You are like a beacon to jihad and you outshine all others.

Al-Aqsa Mosque was engulfed in cheerfulness as the pious fell for its sake and you gave hope to it and it saw all wrongs being righted with determination and, today, it has regretfully been covered in sadness because of the polished dialogues. So hold fast to solid ground and do not be appeased by condemnation or denouncement. He is the God of the heavens and Earth. If it were not for God first, as well as your armed attacks and your legitimate jihad against the forceful occupier and your endurance of pain for the sake of that jihad, as well as your great sacrifices and your attacks on those connected to the damned ones with your successful operations, then you would not have had any weight or clout in this mess and you would have been another number blown away by the storms of conspiracies, a number that is not considered or acknowledged. Instead, when these political leaders reached all that they have reached, they excelled through their elections which are tainted by the blood of the martyrs and they fed off the efforts of the brave and rested on the bodies of the wounded until the moment when their feet touched the grounds of the pagan legal council, taking shelter in its dome and resting on its seats to suddenly revive the laws of Arafat and afflict it upon the broken Palestinian people.

Now they make their constant rounds on all the capitals of deception and malice – Moscow, Cairo, Tripoli, Tehran, Riyadh, Damascus, and others – and go to a series of meetings with the pharaohs and tyrants of these times and the leaders of the nations. They go to the politics of begging, pleading, and concessions, to different organizations and committees, which were only established to support the Zionist state. It is the right of each one of you, the truthful heroes of HAMAS, and the right of each Muslim everywhere, to ask the direct question: What do the political leaders of HAMAS want? And where are they taking it? And it is everyone's right to ask them for an honest answer so as to judge their doings and to evaluate the truth of their claims, away from the trivialities of the statements published in the despicable media and away from the gossip and the empty promises which do not shelter this poor people from the wrath of the bombs of the Jews or from the betrayals of the Palestinian government and its doings. It is impossible for someone who knows the truths of his religion and who is keen on reaping the fruits of his sacrifices to agree to be led by leaders such as these politicians who have proven their failure and the deviousness of their thoughts and the disintegration of their plans, despite what they claim of tact, understanding of politics, maturity and the ability to adjust to the changing times.

By God, a building raised on such foundations will definitely collapse with the placement of the first brick; it will defy its builder and rise to attain its goal, if he honestly wants to reach it, but he will not attain it. Those who build upon a foundation of fear of and approval from God are better off then those who build on shifting ground which eventually will collapse, taking them down to Hell. God does not lead the oppressors to the path of righteousness. Whoever wants to build based upon religion and wants to establish the basics of jihad and reinforce the pillars of truth and reach a clear goal must know the truth of his religion and the ways upon which it is founded. Otherwise, his efforts will continue to be futile. Do you suppose he will hope to reap anything after that? That is impossible because he who wishes to meet his God must do good deeds and must believe only God. Our last supplications are praise be to God, Lord of all creation. Prayers and peace be upon His gentle prophet.

The Tawheed of al-Saud... and the True Tawheed

Release Date: 30 May 2007
Production Date: May/Jun. 2007
Type: Audio statement on video
Version: Transcript

The Tawheed of Al Saud... and the True Tawheed Shaykh Abu Yahya Jumada al-Uulaa 1428

"O you who believe! Take not the Jews and the Christians for your friends and protectors: they are but friends and protectors to each other. And he amongst you that turns to them (for friendship) is of them. Verily Allah guides not a people unjust." (5:51)

All praise is due to Allah Alone, and peace and prayers be upon he after whom there is no Prophet, and on his family and companions.

By the grace of Allah, the Glorious and Great, the Mujahideen in the Arabian Peninsula - may Allah guide them - are people of precedent who have left their good mark and are in no need of someone like me to explain a ruling, dispel a misconception about them, or expose a deception. Ever since they rose up in the face of the tyrants of Al Saud, the ink of their pens has been flowing with living words filled with the spirit of Iman, glowing with the light of knowledge, ruled by the reins of justice and equity, and standing on the foundation of Tawheed and understanding.

The writings, messages and books of the Mujahid Shaykh Yusuf al-Uyayri (may Allah have mercy on him) have accompanied the march of Jihad there step by step, and the words of Shaykh Abdullah al-Rashood (may Allah have mercy on him) have never stopped having an effect on the hearts of the youth and spreading in them zeal, strengthening resolve, reviving hope and dispersing despair. And many are the other scholars of Jihad and seekers of knowledge who have left their mark on the library of Jihad with studies, investigations, rulings and rebuttals which have regulated the march, clarified its boundaries, established its concepts, muted those who impugn its truthfulness, and replied to the fabrications of its enemies. These scholars have stood up, openly declaring the truth and laughing at the strength of falsehood, and called out: "Let us debate, for argument is fought by argument, evidence is met by evidence and investigation is confronted by investigation, not 'investigators.'" But the only argument of bankrupt, fallen falsehood was to follow in the footsteps of their commander and teacher: "He said, 'If you take any god other than me, I will make you of the prisoners!'" (26:29)

And today, the tyrants of Al Saud have resumed the attack. Before the convening of the international conference for the support of Iraq, they wanted to offer a token which would show the earnestness of their attitude in combating what they call "terror", and show that they are still unwaveringly loyal to the covenant. So they waged their violent attack on the Mujahideen in the Arabian Peninsula, and to strengthen the attack, they motivated their information organs to blow and puff until it seemed to the listener that the defenseless people under arrest were the hordes of Genghis Khan which would have devoured everything and destroyed man and stone.

And also, what we are accustomed to with the government of Al Saud is total information blackout and the burying of any news from which it might be understood - even if by analysis - that there is a domestic opposition depriving their ruling regime of sleep. So why today - when conditions in the region are blazing - have they departed from their "wisdom" and rituals and gone out and boasted of their attack and inflated it and publicized it, until the Qibla of their rule, the White House, announced its praise of them and patted their shoulders in expression of its pleasure and gratitude at their actions?

In spite of that, the tyrants of Al Saud didn't stop at this point and didn't suffice with this support, but headed in another direction in which they saw the propping up of their reign and strengthening of their back: i.e., procuring fatwas which aid what they perpetrate and support them in the deeds they commit, for the sleeper to sleep deeper, the confused to increase in confusion, the criminal to become more daring, the repressed Mujahid to increase in repression and rage, and the arrogant tyrant to become even more tyrannical. Thus, among the things which I came across recently was a fatwa published by the Mufti of their kingdom in connection with the attack waged by their security services against the youth of Jihad on the Peninsula in which dozens were arrested according to their information media. And when I saw the fallacies, rashness, misplacing of matters and misapplication of Aayaat and Ahadeeth, it occurred to me to speak these words, as the fatwa in question put it, "To discharge the duty, free myself of responsibility, explain the truth, and out of honesty to Allah, His Book, His Messenger, and the leaders of the Muslims and their rank and file." I ask Allah, the Glorious and Great, to benefit with them their speaker, their listener and their distributor. Verily, He is the Hearer, the Knower.

First: It is very obvious, from the terms repeated in the fatwa, that total reliance in describing the state of the "group" regarding which this fatwa was published was on the communique of the Interior Ministry, as the Mufti said at the outset, "This, and the communique released on the day of Jumu'ah, 10/4/1428 AH, by the Interior MinistryÉetc." and repeated three or four times the phrase, "Among that which appeared in the communiqueÉ" It suffices to topple a fatwa that it relied in "verifying its reason" on a concocted communique released by the Interior Ministry headed by the Imam of the Imams of Kufr, "dean of the Arab interior ministers" Nayef bin Abd al-Aziz, who utilizes the men of his ministry to count the peoples' breaths, follow their every move, violate the sanctuaries of their houses, and torture those of sincere Iman and pure Tawheed who refuse to let their Tawheed be a monstrosity which they are proficient in talking about and with the explanation of which their tongues run as water runs from the mouths of bottles, except that when they turn to the reality of their state and see its monstrous actions which destroy Tawheed from its root, they shut their eyes and plug their ears and say, "We neither saw nor heard," and they swear their greatest oaths by Allah that it is the state of Tawheed and protector of the sanctity of Shari'ah.

The Interior Ministry on whose fabrications the Mufti relies is the same one who stands publicly and proudly and in word and in deed and side by side in aiding the Arab states in their combating of the Muslims. The conferences of the Arab interior ministers continue to be convened time after time to enforce conventions and strengthen ties to combat what they call "terror" - which in their dictionary is nothing but Islam, even if many people refuse to understand or comprehend this: "If Allah decrees misguidance and punishment for someone, you can be of no help to him with Allah." (5:41)

The Interior Ministry whose misguidance the Mufti uses as his source is the same one who lets loose its beasts to sink their claws into the pure bodies to tear them apart out of deep hatred and burning vengeance in the darkness of the prisons. The tongues of these believers constantly remember Allah, exalt Him, seek His aid, and complain to Him, whereas the tongues of those crude executioners scream with the insulting of the Lord, mocking of the religion, and derision of the believers, and vomit ugly, dirty and foul language. This fact has become definite and related over and over again, however much you deny it, or turn away from it. And this matter isn't new, as some think, but rather it became well-known recently, after the area of battle widened and the people were divided into two distinct parties, each of which raised the banner of that which they worship. "And among men are those who take others as rivals to Allah: they love them as they should love Allah. But those of faith are overflowing in their love for Allah. If only the unrighteous could see when they would see the punishment, that to Allah belongs all power, and that Allah is severe in punishment." (2:165)

Shaykh Abu Layth, may Allah preserve him, describing the first moments of his detention in the prisons of the Interior Ministry of Al Saud, says, "To begin, they started torturing me instantly without questions or prefaces. When they discovered my Islamic identity, they wanted to dispel the idea of religious restraint which I might have about them, so the general who was supervising my torture, whose name was Amin Zaqzooq and was of Egyptian origin, began to talk to me and insult the religion. I realized that he wanted to provoke me, so I pretended to be indifferent. Then he followed that up by insulting God, but I pretended not to care. Then he continued by insulting the Messenger of Allah (peace be upon him). So when he found me to be indifferent, he literally told me, "Were Ibn Baz and al-Uthaimeen here - and he mentioned a number of Mashaa'ikh - were they here, I would have ****ed them!" He said that just like that, using the gutter term and meaning sodomy. And after that he started torturing us." From an interview with al-Fajr Magazine.

And if we wanted, we could write for you page after page of these stories, horrors and atrocities, not by concoction and fabrication, but with truthfulness, reliable sources and details of events. And if you insist on turning away, then you are just like those about whom Allah, the Glorious and Great, said, "And if they see all the signs, they will not believe in them; and if they see the way of right guidance, they will not adopt it as the way; and if they see the way of error, that is the way they adopt." (7:146)

And as the whips were tearing apart the bodies of the youth of Islam in the darkness of the prisons of the Interior Ministry, and the mouths of the executioners were spewing blatant Kufr and humiliating insults, the verse of Hirabah (warring) - like today - was being recited, broadcast and printed in the information media of Al Saud's government, to be an unsheathed sword with which the necks of the devout and worshipful are struck. And from who? From people who were expected to declare the truth, restrain the wrong-doing criminals, and rescue the tortured and oppressed who can do nothing but cry, "O my Lord! Build for me, in nearness to you, a mansion in the Garden, and save me from Pharaoh and his doings, and save me from those that do wrong." (66:11)

The Interior Ministry in whose lies the Mufti has confidence and whose rumors the Mufti believes is the same one who mobilized its dogs and sprang to its feet when America became enraged and agitated and its fool said, "Whoever is not with us, is against us." It took it upon itself to be the best of helpers, and its argument to you and in front of you was. "We fear lest a change of fortune bring us disaster." (5:52) And be certain: they "swore their strongest oaths by Allah that they were with you. All that they do will be in vain, and they will become losers." (5:53)

And it performed what was asked of it and more, and so it arrested who it arrested from the scholars who speak the clear knowledge and declare the pure truth, although they were not of those who "prepared for fighting with weapons," nor did they "declare the Muslims to be infidels," nor did they "permit the spilling of their blood," nor did they "rebel against their leader," nor did they "plot to assassinate public figures," nor did they "collude with foreign parties against the country," as the fabricated charges to strike the necks always claim.

So why then are the prisons of the Interior Ministry full of them?

"And they punished them for no other reason than that they believed in Allah, the Strong, the Praiseworthy." (85:8) "And you only wreak your vengeance on us because we believed in the signs of our Lord when they reached us! Our Lord! Pour on us patience and constancy, and take our souls unto you as Muslims." (7:126)

This Ministry has also handed over a number of Mujahideen to the Americans, and at the handover, they saw them off with insults, curses, and slander and strutted and boasted about that in front of their American masters, while the captive Muslim was sad, humiliated, worn-out and naked as he was pulled from the vehicles of the Interior Ministry to the planes of the Cross-worshipers, his heart breaking with grief and sorrow because he doesn't know which prison will swallow him just as he doesn't know which prison spit him out.

And at the same time, this Ministry has opened the doors of its bureaus of investigation wide open to the Romans for them to come and go as they please, and question whomever they want whenever and however they want. And by Allah, it has often been the case that the interrogators of the CIA and FBI were kinder and gentler towards the Mujahideen than were the harsh officers of the Interior Ministry whose sole concern is providing a piece of information to gain the pleasure of the Cross-worshipers, even if they know deep down inside that the information is falser then Musaylimah's "revelation."

This, then, is the Interior Ministry on who details and communique the Mufti depends and makes it his evidence on which he bases his momentous rulings against young men who are cleaner and purer than rainwater.

And one really does pause in confusion and bewilderment as the questions swirl in his puzzled mind: is it really possible that all these things - which have become undeniable even by the elderly Bedouins - are hidden from these people? These stories and reports are recurrent, and investigating and confirming them to be certain about them couldn't be easier, so why this insistence on exonerating these criminals by dredging up arguments which - by Allah - they've never even thought of and which they are incapable of even comprehending, much less relying on?! And why continue to shut the eyes to their flagrant, blatant crimes the denial of which is akin to sophistry in undeniable things? O Allah, show us the truth as truth and grant us following of it, and show us falsehood as falsehood and grant us avoidance of it.

Second: the Mufti opened his fatwa with what the Imam Muhammad bin Abd al-Wahhab (may Allah have mercy on him) mentioned concerning the issues of the Jahiliyyah, and that from their attributes is their disunity in religion. This is a word of truth in the wrong place, and unreliable, exaggerated alarmism about the event in question which only scares those who hang on to the fringes and haven't grabbed the firm handhold nor reached the core of the truth. Otherwise, I think that many of those who have been swallowed by the prisons of the Interior Ministry have memorized by heart the books of the Imam and have understood the knowledge, rules and doctrines contained in them, and have moved according to them, aided them and called to their contents, and have been harmed for applying them.

Thus, when the tyrants of Al Saud realized the extent of these books' influence and that they and their religion are incompatible and that every time they distorted and falsified, they were exposed by them in the worst way, they began to distort the curricula based on them and subject them in a way that runs with their whims and agrees with the demands of their masters. And they muzzled the mouths of the callers who reveal the truth through these living books, books to which the successive governments of Al Saud had long pretended to be related and which - by their claims of concerning themselves with them and printing them - they had deceived the people. But they didn't realize that they were in fact like those about whom Allah said: "They destroy their dwellings by their own hands and the hands of the believers. So consider, oh you with eyes!" (59:2)

So finally, they found themselves at a crossroads with these books: they couldn't twist them to keep them in line with their desires, nor could they follow them, even if only by word. So they chose to be hostile to them and throw them behind their backs. And such is the truth whose people honestly, sincerely and impartially record: falsehood cannot make use of it in aiding its falsehood, and if it tries, it will be exposed sooner or later: "And no question do they bring to you but we reveal to you the truth and the best explanation." (25:33)

So the proper direction for talk to turn to is: is what these young men do from the disunity in religion that Imam Muhammad bin Abd al-Wahhab had in mind, or is it a baseless allegation and reckless throwing about of words without study or reflection?

And what is the religion which these people refuse to let it bring them together with the tyrants of Al Saud and take them under its umbrella and dome? Is it the religion of Islam or the religion of the United Nations? Is it the religion of obedience to the Shari'ah or the religion of ruling according to the Security Council? Is it the religion of purely faith-based brotherhood or the religion of the international community, international family and international legitimacy? Is it the religion of unity of creed or the religion of the League of Arab Countries? Is it the religion of loyalty to Islam or the religion of the Gulf Cooperation states? Is it the religion of disowning the infidels and declaring hostility to them, or is it the religion of brotherliness and affection towards every atheistic arrogant tyrant? Is it the religion of aiding the weak and oppressed and expelling the occupiers, or the religion of surrender initiatives and supporting the infidels against the people of Islam? Is it the religion of combating polytheism and polytheists, or the religion of protecting the criminal Rejectionists and defending them as they insult the Companions of the Chief of the Messengers?

We place these questions in front of the Mufti, and we demand of him frank and daring answers full of proof and detail, and afterwards let him tell us - - with the courage of the man of knowledge - if these are of "those who split up their religion, and become sects, each party rejoicing in that which it has." (30:32) And let him bring for his answers whatever he wants from what was recorded by the Imam Muhammad bin Abd al-Wahhab and the Imams of the [Najdi] Da'wa (may Allah have mercy on them), whose books continue to speak the truth and launch it to disperse the meteorites of falsehood, and from whose books the tyrants of Al Saud would prefer that not even one page remain to make them lose sleep: "When our clear signs are recited to them, you will notice a denial on the faces of the unbelievers! They nearly attack with violence those who recite our signs to them. Say, 'Shall I tell you of something worse than that: the Fire which Allah has promised to the unbelievers! And terrible is that destination!" (22:72)

Imam Muhammad bin Abd al-Wahhab said about the issues of Jahiliyyat that among them is killing those people who command to just dealings, and he was right - may Allah have mercy on him - for Allah, the Glorious, the Great, said, "As to those who deny the signs of Allah, and in defiance of right, slay the Prophets, and slay those who teach just dealing with mankind, announce to them a grievous chastisement." (3:21) And isn't that exactly what the tyrants of Al Saud do? How numerous the callers to just dealing who have been killed, the Imams of guidance who have been tortured, and the Mujahideen who have been displaced, and not even the pure, chaste women have been safe from them, in Pharoanic vileness and villainy, for all to be strung on the thread of "I only show you what I see, and I only guide you to the right path!" (40:29) And their proof with which they legitimize the taking of their blood and honor is: "These are a small band, they have enraged us, and we are all forewarned." (26:54-56)

And among the issues of Jahiliyyat which the Imam - may Allah have mercy on him - listed: their affection for infidelity and infidels. And the state of the tyrants of Al Saud in this regard is clearer than the midday sun, and "nothing is right in the minds when the daytime requires proof." Their crime hasn't stopped at the affection of the heart of which they will be exonerated with heavy oaths and the proffering of minute arguments which they never even thought of. No. their loyalty and aid to all types of infidels and their backing of them against the Muslims is something they boast about in their conferences, meetings and press, and due to its abundance, it has become accepted by the souls of many, and they no longer notice it, much less appreciate its shame and ugliness.

Shaykh Muhammad bin Abd al-Wahhab (may Allah have mercy on him) said in regard to the nullifiers of Islam, "The eighth: supporting the infidels and helping them against the Muslims. The proof is Allah's statement: 'And he amongst you that turns to them [for friendship] is of them. Verily Allah guides not a people unjust.' (5:51)" And Shaykh Hamd bin 'Atiq (may Allah have mercy on him) said, "As for hostility to the infidels and polytheists, know that Allah (the Glorified and Exalted) has made that obligatory and emphasized its obligatory nature, and made loyalty to them illegal and was strict in that, to the extent that there is not in the Book of Allah a ruling whose proofs are as numerous and clear as this ruling, after the obligation of Tawheed and prohibition of its opposite." (Sabeel al-Najaah 31)

So establishing Western bases in the Land of the Two Sanctuaries from which their planes take off, planes which carry tons of explosives to demolish the houses of the Muslims in Iraq and Afghanistan and which go out in the morning with their bellies full and return with them empty: is that backing of the infidels against the Muslims or not?

And evacuating the ports and beaches at which ships drop anchor as they carry thousands of soldiers, dozens of aircraft and hundreds of missiles which pound the Muslim lands and leave them desolate, devastated and flattened: is that backing of the infidels against the Muslims or not? And leaving the coffers of the banks filled to the brim and wide open for the infidels of the West to scoop from it whenever they want and however much they want, for it to be provision and relief for them in their war which has ruined their economy: is that backing of the infidels against the Muslims or not?

And the pumping of millions of barrels of petrol - whether free or for sale - - to irrigate their planes, tanks, ships and cars as they kill, torture and destroy: is that backing of the infidels against the Muslims or not?

And providing the occupation troops with the finest kinds of food and drink for them to enjoy them in front of the lost and hungry prisoner and to strengthen themselves with them in their open war against Islam and its people: is that backing of the infidels against the Muslims or not? So you are caught between two things from which there is no escape: either you deny that these things have been committed by the tyrants of Al Saud, and the first to call you a liar and shout in your face will be them, or you admit it - and you have no other choice - in which case you must reveal their ruling, explain the truth about them and declare your disowning of them and their infidelity. And that is what we hope for.

Third: what we have mentioned negates the argument in support of which the Mufti related a number of proofs which obligate obedience to the rulers, forbid rebellion against them and threaten those who nullify their pledge after they have made it or die without a pledge around their necks, because one of the Ahadeeth which he brought indicts him and makes clear that all of this is out of place. It is the Hadeeth of 'Ubadah bin Samit (with whom Allah was pleased), in which he said, "We pledged to the Messenger of Allah to listen and obey in difficulty and ease and in what we like and dislike and even if we are not given our full due, and that we not challenge the rulers and that we say the truth wherever we are and not fear for Allah's sake the blame of a blamer." And in one version, "and that we not challenge the rulers unless you see blatant infidelity about which you have proof from Allah." (Agreed upon)

The blatant infidelity committed by the government of Al Saud is one of the things that has made the Mujahideen arise and apply themselves to removing it in obedience to Allah, who said, "And never will Allah grant to the unbelievers a way over the believers." (4:141)

And in obedience to the Messenger of Allah, peace be upon him, who said, "Unless you see blatant infidelity about which you have proof from Allah," and to follow the consensus - which more than one scholar has mentioned - about the obligatory nature of removing the infidel ruler and installing a Muslim ruler who is to be listened to and obeyed, and to purify the Arabian Peninsula which the Companions - with whom Allah was pleased - washed with their blood and which the tyrants of Al Saud desecrated by bringing in the armies of infidelity, fornication and corruption from every sect and religion, who established their bases in that blessed land and deployed their troops, and it became a secure refuge in which they shelter and protect themselves after committing the ugliest forms of destruction, killing and expulsion against the Muslims and their lands. The Prophet, peace be upon him, cursed the one who shelters a criminal, so what about the one who encircles the butchers in welcome, receives them with honors, shelters them in his kingdom, gives them aid, and punishes severely those who aim for them and pours torment on them, to please the criminals, gladden the murderers' hearts, and delight the butchers.

So what do you want from these jealous youth following this? Should they be emasculated in their resolve, petrified in their hearts, without concern or zeal, licking the mires of this world, belching from its delights, and competing for its rubble, as their Ummah is felled, wounded and slaughtered, shrieking beside them and imploring its Lord, and wailing and wailing and wailing while they are sitting and giggling and playing? And the Prophet, peace be upon him, said, describing the Muslims, "They are as one against the others." And he said, peace be upon him, "Expel the polytheists from the Arabian Peninsula."

So it was not for the heroes of Islam and the jealous ones among them, as they see their neighbors and brothers in creed and religion having their homes crushed, their countries devastated, their children orphaned, the honor of their women violated, their youth and elderly humiliated, the roots of the religion pulled out of their lands, and the treasures of their countries pilfered, all at the hands of the Cross-worshipers who reside with their deadly weapons and murderous armies in their midst on the Arabian Peninsula in secure, tranquil bases surrounded by the villainous troops of the tyrants of Al Saud: it was not for them to stand like mummies, neither moving a finger, nor fighting a tyrant who supports these murderous criminals and has opened his land to them, mobilized his army to protect them, spent his wealth to strengthen them, and placed for their security his security men who sacrifice their lives for the lives of their brothers the infidels when they should be sacrificing them for the disaster-stricken Muslims in Iraq, Afghanistan and elsewhere. And Jabir - with whom Allah was pleased - related that the Messenger of Allah - peace be upon him - said, "There is no one who abandons a Muslim in a place where his honor is violated and his sanctity is infringed upon except that Allah, the Most High, abandons him in a place in which he would like His aid. And there is no one who aids a Muslim in a place where his honor is violated and his sanctity is infringed upon except that Allah aids him in a place in which he would like His aid." (Abu Dawood)

So this miserable, humiliating situation is what called them to the readying of weapons, which the Mufti found repugnant and in that kept in line with the communique of the Interior Ministry, saying, "And among that which appeared in the communique is their readying of weapons." But what they did is nothing but a response to the divine call, "Against them make ready your strength to the utmost of your power, including steeds of war, to terrify with it the enemies of Allah and your enemies, and others besides them, whom you may not know, but whom Allah knows. Whatever you spend in the cause of Allah shall be repaid unto you, and you shall not be wronged." (8:60) And 'Uqbah bin 'Aamir (with whom Allah was pleased) said, "I heard the Messenger of Allah (peace be upon him) on the pulpit saying, "'Against them make ready your strength to the utmost of your power." Verily, strength is firing; verily, strength is firing; verily, strength is firing." (Muslim)

And then the Interior Ministry's communique made up something which the Mufti followed it in by saying, "ÉAnd their plotting to rebel against the Muslims with those weaponsÉ" This, by Allah, is a lie which of late has been promoted and marketed in all the states which the tyrannical executioners control. So who are these Muslims whom the Mujahideen are accused of plotting to rebel against? If they are their general public, then by Allah, the Mujahideen have only gone out in defense of them and to lift from them oppression, repression and suppression and aid the weak and oppressed moaning under the burden of the laws of infidelity which rule and enslave them. And they are encouraged in that by the statement of Allah, the Glorious and Great: "And what is the matter with you, that you do not fight in the path of Allah and of the weak and oppressed men, women, and children whose cry is, 'Our Lord! Rescue us from this village whose people are oppressors, and raise for us from You one who will protect, and raise for us from You one who will help.'" (4:75)

Otherwise, killing the Muslim public needs no plotting, for the streets are busy and the markets are crowded. But it is the fabrications of the bankrupt and misconceptions of those lacking arguments whose chests are full of hatred. So remember, oh you Mufti, that one day you shall stand in front of Allah the Most High, who said, "Not a word does he utter but there is a vigilant guardian." (50:18) And He said (Glory to Him), "And pursue not that of which you have no knowledge, for surely, the hearing, the sight, and the heart: all of those shall be questioned." (17:36)

If, on the other hand, what is meant by the Muslims against whom rebellion is being plotted are the tyrants of government and protectors of oppression and their parties which refrain from many of the prominent laws of Islam, then what a fine plot and what a fine preparation, and their efforts in that are to be appreciated and their deeds are - with the permission of Allah - righteous. Shaykh al-Islam (may Allah have mercy on him) said, "Every party which departs from any prominent, recurrent law of Islam must be fought compulsorily, by agreement of the Imams of the Muslims, even if they say the two testimonies of faith. So if they agree to the two testimonies but refrain from the five prayers, it is obligatory to fight them until they pray, and if they refrain from the Zakaat, it is obligatory to fight them until they pay the Zakaat. And the same if they refrain from fasting Ramadan or Hajj to the Ancient House, or if they refrain from the prohibition of indecent deeds, adultery and fornication, gambling, alcohol or other prohibitions of the Shari'ah. And the same if they refrain from judging in matters of blood, wealth, honor, sexual relations and the like by the Book and Sunnah." (Majmuu'a al-Fatawa 28:510)

And how strange it is to see the government of Al Saud be described as a "just authority," but the amazement ends if one reads the statement of Allah (the Glorious, the Great): "For any to whom Allah gives not light, there is no light!" (24:40)

I didn't want to go here into all the doors of infidelity from which the tyrants of Al Saud have left Islam, because the efforts of the sincere, impartial, truth-declaring scholars have finished that and explained them nullifier by nullifier in various topics and numerous occasions. Instead, I just wanted to acquaint them with the clear fact which the uneducated recognizes before the scholar, the infidel recognizes before the Muslim, and indeed, the tyrants of Al Saud themselves brag about at every opportunity, which is the issue of the aid, backing and help which I referred to just now, and which we demand of every one who defends this apostate government and hangs on to it by his teeth to disprove with any means at his disposal: whether with the clear and definite legal evidence - which isn't possible - or by bending the necks of the texts or even breaking them. And if they are unable to do that - and they are unable without a doubt - then let them take off the noose of feebleness and cloak of impotence and speak the truth or refrain from aiding falsehood, for there are in the knights of Islam and eminent scholars those who will expose, scatter and spite it:Ê "Those who preach the messages of Allah, and fear Him, and fear none but Allah. And Allah alone is sufficient." (33:39)

O Allah, deal with the tyrants of Al Saud; O Allah, count their number, and kill and eliminate them, and leave not even one of them. O Allah, empower over them your soldiers and your believing slaves, and make them a sign for those who consider, and grant victory to your slaves the Mujahideen and protect them and their families, and defend them, O You who defend the believers.

And our final prayer is that all praise is for Allah, Lord of the worlds.

As-Sahab Media 1428

The Masters of the Martyrs

Release Date: 31 Jul. 2007
Production Date: Jun./Jul. 2007
Type: Video statement
Version: Translation

In the name of God, the Merciful, the Compassionate

All praise is due to God, and prayers and peace be upon the messenger of God, and upon his family, his companions, and all those who followed him.

Nation of Islam,

May the peace, blessings and mercy of God be with you.

Long ago, a poet said, "Killing is destined for us, fighting is for us, and staying behind is for women," but today we are in an era in which the situation is reversed. It is an era in which men are wearing the veils of women, and are staying inside their homes, hiding in the dark corners of their houses, after they held back from entering the areas, the fierceness of which are entered only by heroes. They are paving the way for the weak women, so that they might do something for their nation, which they had refused to do.

Alas, what a pity to see an era in which weakness has settled, and feebleness has nestled among its people, leaving them overpowered by cowardice, and forcing their veiled and anklet-wearing women to take the stand of men, and go to the battlefields at the calls of Takbir, and search for a supporter and for the calls for help. They are moved by living faith, live conscience, and hearts filled with exploding volcanoes of fervency for the truth, and flowing rivers of sorrow over a slandered religion, a violated honor, and an insulted Sharia, in a country that is corrupted by the dishonest, and controlled by the apostate infidels, its womanlike men and worshippers of lust. We all heard about Hafsa University in the city of Islamabad in wounded Pakistan, which took a stand that made it worthy of its name as the university of Hafsa, the daughter of Umar, may God be pleased with both of them. It took a stand in which chastity and virtue screamed in the face of profligacy, shamelessness and debauchery and the voices of pride in faith and in religion were heard decrying and belittling the calls for the sinful modern civilization and the shameless Western freedom, which is promoted by the advocates of depravity in Pakistan to catch up with the procession of modern al-Jahiliyyah.

Faith called loudly until the ground shook by its echo, and the fragile pre-Islamic like entity trembled. "Do they then seek after a judgment of (the days of) ignorance? But who, for a people whose faith is assured, can give better judgment than Allah," [Koranic verse; al-Ma'idah 5:50].

It is true that the university's degree might not have a great weight in the world of worldly degrees, which are pursued by many. But this time, and with its [honorable] stand, its certificate received the highest medals of honor, and earned the highest status, and was recognized, and respected by everyone who loved or hated it, because it is a certificate of righteousness, truth and faith. It is a certificate of righteousness because it uttered the truth, and raised the banner of guidance and echoed with the words of certitude, and supported the calls of purity and shouted in the world of deep darkness "Verily, this is My way, leading straight: follow it: follow not (other) paths: they will scatter you about from His (great) path: thus doth He command you. that ye may be righteous." [Koranic verse: al-An'am 6:153].

It is a certificate of righteousness because it was motivated by the zealousness for the religion, and was moved to rebel by the fervency to support the faith, and was driven by disdain for submission to the promoters of indecency. Despite its weakness, its self-esteem brought it out by its sincere belonging to Islam, and threw in the face of the despicable wrongfulness, "We should indeed invent a lie against Allah, if we returned to your ways after Allah hath rescued us there from" [Koranic verse; al-A'raf 7:89].

It is a certificate of righteousness because it uncovered the deceitful wrongfulness and removed the fake covers from the black grim faces of its owners and forced them out of their deceptive hideaways to stand in front of everyone, stripped, exposed and humiliated, with nothing to cover them, and were thrown disgracefully were they belong, "Those who resist Allah and His Messenger will be among those most humiliated" [Koranic verse; al-Mujadilah 58:20].

It is a certificate of righteousness because it sprouted from the pure instinct, and sprang from the depth of monotheistic hearts, and rose from the moral constitution of the pure souls. It did not wait for permission from a disguised liar, nor a recommendation from a fake adulator; nor a recognition or an appreciation of a ravaging careless tyrant. The permission and the recommendation and the recognition are all in the noble saying of He who has Grace, "Let there arise out of you a band of people inviting to all that is good, enjoining what is right, and forbidding what is wrong: They are the ones to attain felicity" [Koranic verse; al-Imran 3:104].

So, good cheers to that group who sought cover under the shadow of lofty manners and jumped on the summit of honor and answered the righteous call with calmness and confidence in the middle of the noise and racket of the calls of corruption and indecency.

You [addressing a female] answered the call of righteousness without hesitation, and you disobeyed the voice of the unchaste liars; you avoided the manners of the enemy graciously, and you followed the manners that were ordained by your Lord; you gave up their outfits and masquerades and made the Hereafter your ultimate desire. Prosper, oh sister, in seeking gratification, so pray in thanks to the Giver of your blessings.

Today, after the Hafsa University in Islamabad recorded its stand, and rose with it to the throne of glory, its professors and scholars refused that their stand would be other than the stand of their students, whom they had raised on the meanings of faith and instilled in their hearts strong determination. They taught them to seek excellence, and paved for them the paths of sacrifice. They crowned the terraces of glory with a crown that shines over the forehead of history, as they restore by their words and actions a symbol whose depth and meaning can only be understood by those who are characterized by patience, guidance and certitude. "I do not mind when I am killed if I am a Muslim, and my death were for the sake of God; and if God wills, He will bless my body if torn into pieces" [poem].

This is how righteousness and certainty spoke in the past. It refused to back off when matters were related to its faith. This is how truth and conviction should be proclaimed today and every day. A person should refrain from satisfying the unfaithful person with words that his despicable nature would wish to hear. Such were the religious beliefs contained in the battle of the

Lal Mosque, or what is called the Red Mosque, and which was truly red, not red for the sake of embellishment or camouflage, but because it was painted, rather covered, with the blood of the faithful and chaste martyrs. This is how we reckon them, but God is the judge, as they took the stance of unparalleled heroes engraving the pages of history with terrific combat stories that can never be matched. And, as this great mosque used to graduate students and scholars, who appealed for compassion, propagated virtue, and forbade vice, it is today graduating a new class of master scholars, who, after receiving their graduation medals, will be among the ranks of the masters of martyrdom. We do consider them as such, but God is the final judge.

At the top of this regiment of lions is the courageous imam and the active scholar, the martyr, the son of a martyred father, and a martyred mother too, as we consider them as such [but God is the final Judge], Abd-al-Rashid Ghazi, may God bestow His mercy upon them, who spoke the truth at a time of subordination, and exalted faith at a time of submissiveness, and despised he who was evil and proud, who depended on his own strength and relied on his own omnipotence. He would confidently, surely and serenely tell him, "You depend on your audacity and conceit, but I depend on God. Get ye then an agreement about your plan and among your partners, so your plan be on to you dark and dubious. Then pass your sentence on me, and give me no respite" [Koranic verse; Yunis 10:71].

He is the one who said while facing the hardship of the blockade and the menace and of the enemy, "I would rather die than give up anything I call for or submit to captivity." His deed matched his word. "Avoiding death to him was easy, but the bitter taste of observance, and the rough road of morals returned him to it. He kept his feet firm in the swamp of death, and said to it under your toes lies hell" [poem].

Did not our prophet, may the peace of God be upon him, say that the best kind of jihad is telling the truth in the face of an unjust ruler. So what do you say about a word of truth that may be thrown in the face of one of those infidels and unjust? rather, in the face of a country, its army, its intelligence and its security authorities? The bearer of the truth conveyed it openly and honestly, without flattery or adulation. He announced it in their midst and while looking at the swords of injustice and revenge shining brightly before his eyes. He did not fear them, nor did he pay much attention to them. He endured for the sake of the truth until he attained his wish and met his end. He was killed and his mother was also killed alongside of him, may God bestow His mercy upon her. He died trying to shut up all the lying mouths, and silence all the hateful hearts that spread lies and falsehoods, as if he was saying to them: "Perish in your rage. Certainly, Allah knows what is in the breasts" [Koranic verse; al-Imran, 3:119]

He was an example for those who fought under him, and became a school for those that will follow his path, God permitting, so that their heads remain high and so that the goal they would compete for is the attainment of the highest rank of martyrdom and its most noble status. "If you take risk for the sake of an honorable objective, do not be content with anything less than the stars. Meeting an honorable death is not more painful than a mundane death" [poem].

Jabir, may God be pleased with him, narrated that the Prophet, may peace be upon him, said: "The master of all martyrs is Hamzah bin Abd-al-Muttalib and any man who was killed because he stood up to an unjust Imam and enjoined for what is right and forbade what is wrong" [Hadith]. Have the martyrs of the [Red] mosque not satisfied this criterion the Prophet used to describe the master of all martyrs? Isn't it true that they stood up to a despotic, secular, and despicable tyrant after they saw him dragging the land and the people to the abyss of abandoning Islam and morality and turning them into blind followers and an exact copy of his masters in the East and West in culture, values, creed, and customs? They stood up to him [Pervez Musharraf] and to his agent and dishonorable army and his Intelligence apparatus, that targets only the weak, in order to say stop this train of corruption that destroyed the country, and crushed the values and obliterated virtue. They stood up to them to say that Pakistan will not achieve independence from the worshippers of cows [India] only to be enslaved by the worshippers of [material] pleasures and the people of sin, "they are only like cattle, nay, they are even farther astray from the Path. [even worse than cattle]."[Koranic verse; al-Furqan 24:44]

They stood up to them to say that Pakistan was not created to be turned into a state that rejects Islam, fights its people, and condemns its commands, and adopts the trash of ideas produced by the minds that were cursed by God and are subject to His wrath and made out of them monkeys and swine, for they worshipped false Gods, and put them in high esteem, and turned them into the laws of the land and forced the people to conform to them in the name of civilization, and progress. They stood up to them to say that Pakistan was not created to become an ally and a supporter of the patron of the Cross, America and its followers, by chasing the mujahidin, detaining them, torturing the monotheists, and opening its airspace, ports, and land to those infidels, and allowing them to move freely in safety while their weapons massacre thousands of Muslims here in Afghanistan. They stood up to them to say that the mission of Pakistan's army, which falsely carries the slogan of "Faith, Righteousness, Jihad for the sake of God," is not to defend the Crusaders, carry out their orders without revision or objection, wreck the mosques, besiege the schools or to kill the people in the streets. But the Pakistan Army's mission is to truly commit, without charlatanism or deception, to the slogan it claims it adopts and carries.

They confronted them to tell them that our Muslim people in Pakistan, the people of virtue, compliance, purity, humility, and zeal, have no room for the people of prostitution, impudence, libertinism and lewdness, who would like for indecency to spread among the believers. They confronted them to tell them that Pakistan is a Muslim nation and its people are Muslims, and it should be ruled by Islam, enjoy [life] under the umbrella of its just Sharia, to have the banner of monotheism fluttering in its skies, and the banners of secularism and the cross rolling in its mud; and if it is not, then below the ground is better than above it. Because of that, they rose, resisted and were killed, and they have become truly martyrs as we count them and God is the final Judge. "Among the Believers are men who have been true to their covenant with Allah. of them some have completed their vow (to the extreme), and some (still) wait: but they have never changed (their determination) in the least: [Koranic verse; al-Ahzab 33:23].

It was as if they were offering their lives one by one, and plunging into the sea of death one after the other, and finding excuses for themselves before God, as the fragrance of faith and longing for paradise is going through their emotions, they repeat in the voice of a man leaving his will: "Brother if you shed tears over me and wet my grave with them in reverence, use them to light the candles' residues, and to guide you toward a time-honored glory; nothing is more generous, no better gift to be bestowed upon and no liberality is greater" [poem].

God willing, their pure and chaste blood that they offered freely, will be prominent landmarks to guide the ramblers, and a flowing spring to irrigate the tree of Islam in Pakistan. The banner that they were loyal to, will be received by those who follow their footsteps, follow their example and imitate their actions, to finish building an edifice that they established with their remains. This is how the procession of truth will continue.

"If one moon [light] is abating or fading, another one will show on the horizon shining" [poem].

A situation like this, which can not be overstated, should not just disappear with the wind and melt in the oceans of lies. This is God's Sunnah. "And say not of those who are slain in the way of Allah. 'They are dead.' Nay, they are living, though ye perceive (it) not" [Koranic verse; al-Baqarah 2:154]

"Oh Martyr with whom God forever carried the battle of justice, you will always be a banner, a leader of the symbol and a troop for sacrifices. We did not forget you have taught us the smile of the believer in the face of death" [poem].

Oh people of jihad in Pakistan, oh lions of sacrifice, and oh you seekers of martyrdom and the virgins of paradise, man dies but once. You should then seek the meadows of martyrdom, which are now available in your country; rise in a way that God will see in you what pleases Him; mobilize to remove this apostate and corrupting tyrant; put an end to his secular tyrannical rule; and, attack the strongholds of his demeaned army, the dens of his filthy Intelligence, and the centers of his pre-Islamic-like regime. You should follow the example of your neighbors, the heroic people of Afghanistan, who, with their steadfastness, strong will, persistence, and with their strong dependence on God have made their country a grave for arrogant and powerful empires and forced them out humiliated and defeated. Along with those empires, the people of Afghanistan crushed all the dishonored agents, and you should do likewise.

You should know that the tax of humiliation that will be paid by the people of Pakistan by their surrender, subordination and humiliation at the hands of this apostate government is many times greater than the tax of glory, which they will pay generously and willingly to glorify the religion, to instill the Sharia, protect the faith, and liberate themselves from worshiping human beings and delight in worshiping the lord of all beings.

Justice will not be won by begging, rights are not attained by pleading, and oppression will not be lifted by imploring, but with tough determination, high-minded aims, serious work, and continuous sacrifices, by overcoming difficulties and with the willingness to take risks. "The greater the peoples' determinations are, so will be the rewards, and the pathetic people will not be able to achieve the smallest of things, but the great people will achieve great things" [poem]. Best of all and more beneficial to you is the Almighty God's saying "Go ye forth, (whether equipped) lightly or heavily, and strive and struggle, with your goods and your persons, in the cause of Allah. That is best for you, if ye (but) knew." [Koranic verse; al-Tawbah 9:41]

We plead with God to accept the martyrs, elevate their rank in paradise, free the prisoners and the subjugated, care for the wounded, cure the sick, grant the families of all those serenity and reward them. He is benevolent and generous. And our final supplication is praise be to God, Lord of all creation.

Dots on the Letters, as-Sahab's 2nd Interview with Sheikh Abu Yahya al-Libi, Two Years After His Deliverance from Bagram Prison

Release Date: 9 Sep. 2007
Production Date: Aug./Sep. 2007
Type: Video statement
Version: Subtitles

Interviewer: All praise is due to Allah, Lord of the worlds, and peace and prayers be upon the Chief of the Messengers, our Chief Muhammad, and on his family and companions.

As-Sahab Media is pleased to again welcome Shaykh Abu Yahya al-Libi, in its second interview with him since his deliverance from Bagram Prison 2 years ago.

To begin, we ask him: Honorable Shaykh, after the passing of this period since your deliverance from captivity and after rubbing shoulders with the Mujahideen and living among them, what is your evaluation of the course of Jihad in general and Afghanistan in particular?

al-Libi: In the name of Allah, the most Compassionate, the Most Merciful. All praise is due to Allah, and prayers and peace be upon the Messenger of Allah and on his family, companions and allies.

Before answering, I thank As-Sahab Media for the major effort it is making to spread Islam's true methodology and clear picture far removed from the patching and fabrication and methodologies of adaptation and compromise whose adoption has become one of the characteristics of the age and perhaps something to boast about.

As for the answer, my noble brother, it can be divided into two parts. The first deals with the scientific/methodological/ideological progress - as it were - which the Mujahideen have achieved and continue to achieve day after day, whether from the perspective of those who carry the methodology inside the Islamic

Ummah to which they belong, or from their perspective of the imposition and spreading of their vision throughout all nations and peoples and their entrance into the field of the battle of ideas and clash of methodologies with strength, confidence, insight and understanding, for the Jihadist methodology to say in the midst of this tumult, "Here I am, so where's the challenger?"

By the grace of Allah, the Glorious and Great, we notice with great clarity a distinction and progress being achieved by the Jihadist methodology in all theoretical scientific directions, through its legal foundations, its analyses of the state of affairs, its outlook on events, its depiction of the issues and its according of specific and clear legal visualizations to many of the major issues which affect the Islamic Ummah at home and abroad.

And we have begun to see and feel the Islamic Ummah's anticipation of the position of the Mujahideen in most of the events which take place today, and the Jihadist voice has become its preferred voice and the one which expresses its opinion and outlook. And the Jihad as a comprehensive methodology is no longer buried among the rubble of ideas or hidden under various other methodologies.

No, by the grace of Allah first and last, it has begun to confront all the methodologies, ideas and deviant concepts; to meet argument with argument and eloquence with eloquence; to discuss treatises and rebut misconceptions; to remove misguidance, correct mistakes and rectify deviation; and to deliver its speech with clarity and pureness and without fearing in that the blame of a blamer.

And the effects of the well-founded scientific presentation of the methodology of Jihad have begun to appear on many of the Islamic movements whose members used to be like those on drugs with concepts and ideas bearing mere resemblance to hallucinations than to anything connected with Law and the state of affairs.

This is as far as the Islamic Ummah is concerned.

If, however, we were to expand the field of sight and move on to the extent of the reach and effect of the voice of the Jihad throughout the world, whether at the level of the governments or the peoples, then we would hear the echoes of this voice resonating from the depths of the infidel world, in Europe and elsewhere.

And as the saying goes, for every action there is a reaction. So this fierce information attack which the people of infidelity are waging in various and disparate forms against the Jihad and Mujahideen is only because of the earthquake caused by the way the Mujahideen analyze and treat the major issues which there [infidels] refuse to look at in a proper, logical fashion far removed from the deception, deceit, distortion and falsification.

And in general, the Jihad, as an Islamic worship and rite which has its rules, manners, regulations and fundamentals, has begin to spread, expand, increase, strengthen and develop, and thus the attempt of the infidel states and behind them their agents in the region to kill the Jihadi spirit in the Ummah, reverse its progress and strangle it with various ropes of constriction is nothing but a kind of futile desperation, and all those efforts they make and that money they spend will be a curse for them and a cause for regret, and the Jihad will continue to prosper, climb and expand, in spite of their dislike of it, in corroboration of Allah's statement, "They seek to extinguish the light of Allah with their mouths; but Allah refuses but to perfect His light, though the disbelievers may resent it" (9:32), and of His statement, "Verily, those who disbelieve spend their wealth to turn people away from the way of Allah. They will surely continue to spend it; but then shall it become a source of regret for them, and then shall they be overcome. And the disbelievers shall be gather unto Hell." (8:36)

As for the second of the two parts, it concerns the practical achievement represented by the armed Jihad being waged today by the Islamic Ummah in many arenas, in which the Mujahideen - who are part of our Ummah - have proven that challenging the enemies in the arenas of combat is a hope which they have long wished for and shown eagerness towards because they knew that the breakdown of the enemy military and the dispersal of his strength will mean without a doubt his decline ideologically, as it is in people's nature to imitate the stronger one, and the battle which they are waging is against the strongest world power. America, which is one the major evil spirits of the age, was only a few years ago bragging about its power and boasting of its army and materiel, at a time when everyone was subordinate to it and submissive to its resolutions: so no criticism, objection or review, but rather, begging, supplication and kissing the steps of the White House and the shoes of its politicians.

But today, where is America. And where is its strength and sham splendor? Where are the slogans of "whoever is not with us is against us"? Where are the threats and warnings? Where is the vanity and arrogance of the American army and its police-makers? And moreover, where is the value of the American soldier, whose killing used to make headlines in all the media, but who today is dragged in the streets of Baghdad, hung on the bridges of Fallujah, rolled on the rocks of Afghanistan and burned to coals in the middle of its capital, Kabul, yet his news passes quickly without interest or analysis, assuming that it was even mentioned by the media and not considered by them to be marginal news for which there was no room in their newscasts or programs.

And here we must ask a frank question: who is it that sent the American forces into this abyss? Who is it that distinguished between the meat and the tumor, and put America on the balance and showed everyone its real weight and value?

You will doubtlessly reply to me - after the grace and guidance of Allah, the Great and Glorious - that they are indisputably the Mujahideen, whether in Afghanistan, Iraq and Somalia or in the heart of America.

If the military gains of the Jihad and Mujahideen were limited to this, it would be enough for them as a source of pride and as a victory. So how about when the matter is much bigger that that and the gains - by the grace of Allah and His aid - are continuing one after the after, and the direction of the battle - Allah wiling - going according to the plans and wishes of the Mujahideen?

And Afghanistan is one of the rings of these confrontations, and in fact, is the mother arena in terms of seniority, and its heroic people have become experienced in confronting the forces of Kufr and are well-versed in tearing apart empires, one after the other, of which we ask Allah that America be the latest of them.

So it is possible to compare the previous state of the Mujahideen from our brothers the Taliban and their helpers with their present state.

The first year after the fall of the Islamic Emirate of Afghanistan - may Allah restore it - was a year of despair and depression for everyone except those whom Allah steadied with the light of conviction, power of faith and confidence in Allah's promise. As for today - and the credit is Allah's alone - the Mujahideen have become the pursuers, not the pursued, and for the most part, the strikers, not the struck. The various types of military operations have begun to strike the depths of Afghan cities, and moreover, the hearts of fortified military bases.

The Mujahideen have taken control of huge, wide swathes of the land of Afghanistan and these areas are now under their authority.

There is now a new generation of Mujahideen pouring by the thousands onto the battlefields, after they discovered the falseness of the propaganda and were delivered from the alarmism which was one of the weapons which the enemy used against them at the beginning of the battle. And the enemy has begun to bumble in his decisions and hesitate even in the resupplying and reinforcement of his forces.

And all praise is for Allah, Lord of the worlds.

Interviewer: OK, you have painted for us a bright picture of the Jihad and Mujahideen ideologically and militarily, but there are those who differ with you in this assessment and believe that the picture is somewhat exaggerated or that the truth is perhaps the opposite of what you stated.

On the military front, for example, many of the Mujahideen's commanders and scholars have been arrested or killed, while on the ideological front, you aren't unaware of the retractions which have been published or attributed to some of the Jihadist symbols and even some of the Jihadist groups. What do you say to that?

al-Libi: It goes without saying that when we say we are in Jihad, combat and defense, this of necessity means that there are sacrifices which there Mujahideen must make, and when we describe ourselves as being in a battle and on differing and open fronts, that means that the battle needs fuel, provision and energy with which to move its wheel. And I didn't say in my previous answer that we haven't offered any sacrifices in the ferocious and heated confrontation between us and our enemies, the Crusaders and the underlings. We have never said this and we can't possibly say this.

But is Jihad anything other than wounds and injuries, killing and combat, and war with its ups and downs?

Everyone whom Allah has guided to observe the worship of Jihad has by that paved the way for himself with a general rule in which he deals with every step he takes in his Jihadi journey: that rule which the Quran had clarified with utmost clarity, saying, "Say, 'You do not await for us aught save one of the two good things'" (9:52), which is victory or martyrdom.

And He, the Great and Glorious, said, "Let then those fight in the cause of Allah who would sell the present life for the Hereafter. And whoso fights in the cause of Allah, be he slain or be he victorious, We shall soon give him a great reward." (4:74)

So the balance and measure by which Islam assesses things is not a purely worldly balance disconnected from the hereafter. The greater part of its scale with which we weigh the value of Jihadi actions is connected to the world of the hereafter, the world of recompense, rewards and the greatest prize from the Generous Giver.

As for the remaining part of this scale, it is the part connected with the world and its interests, and the part in which the sweetness of victory and empowerment is placed. And accordingly, it is with this proper visualization of how to deal with profit and loss, victory and defeat, and success and failure that we must look at our Jihadi actions.

Consider with me the Hadeeth related by Imam Muslim and others from Abdullah bin 'Amr bin al-'Aas (with whom Allah was pleased) who said, "The Messenger of Allah (peace and blessings of Allah be upon him) said, 'There is no raiding party or squadron which raids, takes booty and returns safely that won't have taken two-thirds of their reward in advance. And there is no raiding party or squadron which fails and is afflicted that doesn't receive their full reward.'"

This is why Allah (the Glorious and Great) has forbade the believers to deal with the affairs of the Jihad in this fashion, and has replied to those who think that the path of prudence is by seeking out safety, distancing themselves from risks and withdrawing, saying, "O you who believe! Be not like those who have disbelieved and say of their brethren when they travel in the land or go forth to war, 'Had they been with us, they would not have died or been slain.' They say this so that Allah may make it a cause of regret in their heats. And Allah gives life and causes death and Allah is mindful of what you do." (3:156)

And He (the Exalted) also said, "It is these who said to their brethren, while they themselves remained behind, 'If they had obeyed us, they would not have been slain.' Say, "Then avert death from yourselves, if you are truthful.'" (3:168)

And because of this, Allah (the Glorious and Great) severely reprimanded the group which attempted to abandon the battle and lay down their swords at the mere rumor during the Uhud expedition of the killing of the Prophet (peace be upon him),which is without exception the greatest misfortune the Muslims could possibly be afflicted with until the Day of Resurrection, and he explained to them that the death or killing of the Messenger of Allah (peace be upon him) and his absence doesn't mean going back, nor does it mean loss on the balance of Jihad, nor is turning on one's heels acceptable because of it. He (the Exalted) said, "And Muhammad is but a Messenger. Verily all Messengers have passed away before him. If then he dies or is slain, will you turn back on your heels? And he who turns back on his heels shall not harm Allah at all. And Allah will certainly reward the grateful." (3:144)

Take, for example, the incident of the Red Mosque in Islamabad, which I consider to be a major step forward in contemporary Jihadi life - by all standards. If we wanted to place this incident on the scale of the balance of worldly gain and loss, which its proponents often camouflage with labels like prudence, profound understanding and dealing with reality with wisdom and composure, then its heroes would no doubt be described as reckless, foolhardy and inexperienced individuals who embroiled themselves in a battle they couldn't win, and thus killed themselves and caused the wasting of promising efforts which were in the interest of PakistanÉand so on.

But if we were to put this event on the proper Islamic balance which is tied to the world of the hereafter you will find that something similar occurred in the era of the Prophet (peace and prayers of Allah be upon him), when 70 of the 'Ulama [men of knowledge] of the Companions were killed at the Well of Ma'oonah. And pay attention to our statement, "the 'Ulama of the Companions": i.e., in familiar contemporary terms, the cadre of the state, and the highest-ranking ones as well. IF this were to happen in our era, it would be considered a disaster, policies would be changed because of it, retractions would be made, and bitter criticism would continue from all sides.

But look at what was said by the Companions (with whom Allah was pleased): they considered it to be one of the virtues of the Helpers, as Qatadah said, "We don't know of a neighborhood of the Arabs with more martyrs more honorable on the Day of the Resurrection that the Helpers. Anas said, '70 of them were killed on the day of Uhud, 70 on the day of the Well of Ma'oonah, and 70 on the day of Yamamah during the reign of Abu Bakr.'"

And even greater than that is to receive the pleasure of Allah, as Anas (with whom Allah was pleased) said, "A verse of the Quran was reveals concerning those who were killed at the Well of Ma'oonah, which we read, then it was abrogated: 'Inform our people that we have met our Lord, and He was pleased with us and we with Him.'"

So this eternity-related part is the part lost in the equation of appraising gain and loss in our battle with our enemies.

Of course, this doesn't mean - neither rationally nor legally - neglecting to take precautions; being negligent in taking those which can be taken; not making an effort to correct flaws, remove deficiencies and benefit from experiences and examine them; and not taking the negligent to account. But this is one thing, whereas turning the Mujahideen's sacrifices - however costly they might be - into obstacles and barriers with which we prevent the continuation of the march is something else altogether.

Therefore, the killing of those leaders of the Mujahideen who have had the greatest of impacts on Islam's contemporary battle between the party of the Most Merciful and the party of Satan is part of the sacrifices whose makers were the people most eager for and desirous of them and most appreciative of their value. And even if the killing of these great leaders has some negative effects on the Jihadi groups, these negative effects are nearly equaled or perhaps overshadowed by the positive aspects, among the greatest of which is the proving of the strength of the loyalty of the people of this religion to their Creed and Law, and that they are prepared - in the interest of realizing, establishing and empowering [the Law] - to give up everything, even their souls and lives.

- From those positive aspects as well is the definite proof that our Shari'ah and our Jihad in particular isn't bound to nay individual, however high his value and apparent his impact. No, it is a Creed here to stay and a Law preserved which is further strengthened and empowered in proportion to the blood its people expend for it.

Doesn't the Quran say, "And many a Prophet there had been beside whom fought numerous companies of his followers. They slackened not for aught that befell them in the way of Allah nor did they weaken, nor did they humiliate themselves before the enemy. An Allah loves the steadfast" (3:146)?

And let me give you some examples of that.

Before the killing of the Mujahid Shaykh Abu Musab al-Zarqawi (may Allah have mercy on him), who is more or less the most prominent of the leaders of the Jihad which the area has lost, the American media machine was trying to convince its miserable people that its victory in Iraq will be achieved with the mere killing or capture of this champion, to the extent that it turned the issue of Iraq into the story of Abu Musab (may Allah have mercy on him), and portrayed him as being like the string of the rosary, whose cutting causes its beads to fall one after the other.

But did the dreams and fantasies of the Bush Administration in this regard come true? The answer is in what we see today in Iraq and the major and continuous progress being achieved by the Mujahideen there and the daily heavy losses being taken by the Americans and their underlings.

And similar to that is the martyrdom in Afghanistan of Commander Dadullah (may Allah have mercy on him): the media inflated the issue of his killing and considered it to be a turning point in the course of the Afghan Jihad, claiming that his martyrdom will lead to the breakdown of or a decline in Jihadi activities, especially martyrdom operations. But the daily reality and what is being suffered by the Crusader forces and the government of treason in Kabul clearly gives the lie to those claims.

Today, martyrdom operations strike in the heart of the Afghan capital, Kabul, in Kandahar, Khost and indeed, in all Afghan provinces, and no effect has been seen on the Jihadi field work in Afghanistan, and for Allah is all praise. And this is why the Americans themselves have begun to realize that the movement of the Jihadi march doesn't hinge on the presence of any single commander and cannot fall apart in his absence, and their dealings with the Jihadist groups have started to be based on this conviction, and thus they have gone an knocked on other doors in the hope of winning this round of the match: for example, what they call the "battle of ideas" and so on.

So to summarize this issue, we say: yes, the Mujahideen have presented - with pride and honor - a number of their heroic commanders. They presented Khalid al-Shaykh, they presented Abu Anas al-Shami, they presented Abu Musab al-Zarqawi, they presented Abu Umar al-Sayf and before him Khattab, they presented Mullah Dadullah, they presented Abd al-Aziz al-Muqrin, and most recently, they presented Shaykh Abdul Rashid Ghazi. May Allah have mercy on them all. And they don't hide this, nor do they consider this - in accordance with the balance of Shari'ah - a loss because of which they must stop their Jihadi work. Rather, they consider the blood of these leaders to be an instigator and motivator for them to stay firmly on their path, follow them and make every effort to avenge them. And the Ummah is full of champions who will fill these gaps, and just as the school of Jihad produced them, it will produce others, and just as they have led, others were lead, with Allah's permission.

If one of our chiefs passes, another arises in his stead He says and does what the honorable said and did

As for the story of the retractions, which some are attempting to portray as an ideological defeat for the Jihadist methodology and a crushing deathblow for it, it is one of the rings of the conflict created by the leaders of infidelity in the halls of the Egyptian security organs, who have advertised it as a fine new prescription who might help in treating the serious plight in which they and their masters find themselves as a result of the Jihadi tide flooding the region and world. And some of the Arab states have rushed to snap it up, like the governments of Al Saud, Libya, Jordan and others.

So this is why we don't look at the retractions which have been published or might be published here and there in a restricted fashion, i.e. that so-and-so has abandoned the Jihadist/fighting methodology which he used to carry, lest we fall with them into the merry-go-round of detailed rebuttals and debates, unless it was to remove a misunderstanding, not to debate a fundamental.

Instead, we look at the issue of retractions as being a completely new idea which is part of the system of the battle of ideas which is one of the fierce fronts of the confrontation between us and our enemies, the Crusaders and their underlings. So let us deal with this phenomenon on this basis. Otherwise, we safeguard for those to whom the retractions are attributed their precedent, their Jihad, their status and their worth, and we also take account of the circumstances of many of them in the unusual ideas which emanated from them or might emanate from them, ideas which oozed from the darkness of the prison cells and under the whips of the lashers and the policy of repression and compulsion.

And perhaps you will agree with me that with ideas produced in circumstances like this, you cannot attribute true conviction to their producer.

Interviewer: So what do you consider to be the proper way to deal with this issue?

al-Libi: It is possible for me to summarize for you the way - as I see it - to deal with this critical issue in several points.

First: the conditions being experienced by these captive brothers to whom the retractions are attributed are conditions of compulsion and suppression, extraction of statements by force and pressuring and blackmailing them with dirty methods to lend a foundation to ideas and methodologies which anyone with the least bit of understanding realizes couldn't be farther from having a connection to legal evidence and scientific foundation.

Thus, fairness demands of us that we refrain from considering these newly-proposed ideas and methodologies as being in line with the convictions of their proposers until they speak them and adopt them in complete freedom and of their own accord.

If Law has permitted the Muslim to speak the word of infidelity - which is the most enormous thing that can be said - in a state of compulsion with the serenity of the heart with faith, then what about saying something less than that?

Allah says, "Whoso disbelieves in Allah after he has believed - save him who is forced to make a declaration of disbelief while his heart finds peace in faith - but such as open their breast to disbelief, on them is Allah's wrath; and for them is decreed a severe punishment." (16:106)

Second: we tell those who want to turn the issue of the retractions into a conspicuous title reading, "Here are your brothers, so be like them": it is imperative that a distinction be made between [on one hand] deriving lessons from experience and taking account of what has happened and [on the other hand] turning these experiences into the just judge and final word in issues of dispute and matters of disagreement, because judgment, decision and submission of disputes is only for the Book of Allah and the Sunnah of His Prophet (peace and blessings of Allah be upon him), and with them alone the statements and actions of the slaves are weighed and to them alone their disputes are submitted.

Allah says, "O you who believe! Obey Allah, and obey His Messenger and those who are in authority among you. And if you differ in anything, refer it to Allah and His Messenger, if you are believers in Allah and the Last Day. This is best and most commendable in the end." (4:59)

Thus, it is never right for us to consider the retractions of the retracting ones and the experiences of the experimenters - whatever their rank - to be legal evidence for us to refer to in disputes.

So if someone comes to us and says, "You're still insisting on your way and methodology and holding on to your ideas although so-and-so and so-and-so have gone back on them, and they're who they are in knowledge and precedent," we tell them: yes, they're from the people of knowledge and precedent, but that doesn't accord their statements a sacredness which makes them indisputable, with no room for critique, refutation and objection. Our legal and religious obligation is to measure the statements of everyone who has retracted with the minute legal balance which doesn't oppress in the least, and to gauge those unusual ideas by the evidence, to see the extent of its nearness or farness from the truth. And then - and only then - we can rule any idea emanating from here or there as being mistaken or correct.

As for the total acceptance, complete surrender and random ruling by experiences and retractions without referring them to the Book of Allah and the Sunnah of His Prophet (peace be upon him), this is a slippery slope and pitfall with which one's religion is not safe and which can't possibly be the path of one who seeks the truth and aspires to know and follow it.

Third: we ask Allah to steady us on the truth and steady our Muslim brothers inside the prisons and outside of them.

Even if we suppose that those retractions which are attributed to some emanated from them of their own free will and conviction, then the reasons for backtracking are not always limited to going from blatant error to blatant truth, so that every backtracker can be made a role model for those behind him.

No there is also the changing of hearts and their abandoning of guidance.

Don't we read in the Book of Allah 9the Glorious and Great) the supplication of the people of fear, apprehension and anxiety: "Our Lord, cause not our hearts to become perverse after You have guided us, and bestow on us mercy from Yourself; surely, You are the Great Bestower" (3:8)?

And the supplication most often made by the Prophet (peace be upon him) was, "O Charger of hearts, steady my heart on Your religion."

He was asked about that, and his reply was, "There is no human whose heart is not between two Fingers of Allah's Fingers, and he sets right whomever He wills and causes whomever He wills to deviate."

And it is narrated that Abdullah bin Mas'ud, with whom Allah was pleased, said, "Whoever follows someone should follow one who has died, for the living is not safe form temptation."

We ask Allah to protect us and our Muslim brothers from temptations, open and hidden.

Interviewer: You indicated previously that the issue of retractions is part of the system of the war of ideas being waged against the Mujahideen.

Can you elaborate on the most important pivots of this war and some of the techniques being used in it?

al-Libi: The Crusader, let by America and its underlings, have come to realize that the Jihad is not what they had imagined. It's not merely short-lived military operations which are no more than a reaction to the bitter reality which the Mujahideen see in their Ummah, nor is it merely an attempt to lift economic, social or political suffering, in which case it would be possible to absorb that agitation with some reformist patching to anesthetize the Ummah before returning to performing the surgical operations with which its limbs are amputated without it feeling a thing.

No! And since they have arrived at this truth, they have become active in opening new fronts against the Mujahideen, fronts which employ two pivots.

The first is the pivot of internal disassembling of the Jihadist groups and indeed, the entire Jihadist methodology. After their coming into contact with the Mujahideen and their becoming privy to many of the details of their ideas, which they arrived at through the Mujahideen's literature and communiques or through the discussions which place now and then inside the interrogation centers behind bars, they realized that the matter is larger and deeper than being just atmospheric crackle, temporary reactions or mere reflections of interlinked suffering. They realized that the greatest part of the battle lies in the well-founded convictions and doctrinal-methodological bases which the Mujahideen adopt and through which they work and which represent their real motivational and mover in the actions they carry out against these infidel states and their allies.

And thus, they reflected and calculated, then looked around, then reflected and calculated, then came to the conclusion that a large part of the battle depends on shaking the convictions on which the Mujahideen build their march and cast doubt on the doctrinal bases they consider indisputable, in which case there will occur a splitting or perhaps collapse of the basic foundations and fundamentals on which the Jihadi methodology stands. And we know that practical vacillation, hesitation and confusion is a reflection and expression of doctrinal and methodological vacillation, detachment and darkness.

So this is the first pivot from which is launched the concept of the war of ideas directed against the Mujahideen, which has become an intrinsic part of the sweeping Crusader battle.

As for the second pivot, it is the attempt to isolate the Mujahideen from the Ummah and besiege them in its midst, and consider them to be an alien body growing inside Islamic society which must be removed, because we know that the Mujahideen are no more separable than an inseparable part of the Islamic Ummah in terms of religion, doctrine and affiliation.

The continuation of this visualization and the practical issues which stem from it means the continuation of the flooding of the Jihadist tide and its resurgence through the Ummah's embracing of it and its feeling and conviction that [this tide] is an extension of its efforts and backing physically, morally and economically.

Thus, the Crusaders want to put in place obstacles and barriers which would come between the Muslim peoples and this understanding and feeling, so that all the issues which the Mujahideen raise would no longer be an expression of the conscience of the Ummah, the creed of the Ummah and the outlook of the Ummah, but instead would be deviant, outcast ideas confined to a small group which would act in random and impulsive fashion, in which case the Mujahideen would be hemmed in and start to erode from the inside, and it wouldn't be long before they would fade out and come to and end.

Interviewer: And according to your outlook, what are the methods which the Crusaders might use to achieve this goal?

al-Libi: First, I say - and with full faith and confidence - that this goal will never be fully and comprehensively achieved, and we are convinced of that and haven't the least doubt about it, because what are called convictions or bases are in fact legal fundamentals and foundations built on pure, shining legal evidence which is part of the Islamic religion which Allah has guaranteed to preserve and protect, even if the infidels hate that. And they are also the fundamentals of the victorious group manifest against its enemy until the Last Hour.

Yes, this ideological war might have an effect on some individuals and perhaps groups, and might cause some confusion and disarray in one place or another, but that it might lead to the total annihilation of the Jihadist methodology and its permanent killing: this will never happen, Allah willing.

And to return to your question about the methods our enemies use in their ideological war, I say: these enemies have no morals at which they might stop in their war, and therefore, they don't have - in their attempt to attain victory - a single method to which they stick and don't overstep.

Thus, lying, fabrication, spreading rumors and committing the dirtiest and most despicable of acts are for them all methods completely inseparable form their war. But when talking about the ideological war, it seems to me that [the methods] can be listed in a number of fundamental points.

First, announcing the backtracking of some of the Mujahideen's leaders in prison, their deeming themselves mistaken in what they used to believe and do and their advising their brothers to abandon the path which they are on.

The media is present in force in this operation to run interviews, print articles and books of the retractions and blow them up out of all proportion and portray them as axioms which aren't open to discussion and giving and taking.

And I spoke about this issue previously and mentioned to you the proper bases on which to deal with it.

Second, fabricating some repulsive lies or exaggerating and blowing up some mistakes from which no field of Jihad is free and considering them to be deviations glued to the Jihadist methodology and an inseparable part of it, and widening their circle to make them a general rule covering all Jihadi groups in all fields of Jihad.

Interviewer: Could you give us an example, Shaykh?

al-Libi: Like the allegation that the Mujahideen deem the Ummah and its 'Ulama to be infidels and legalize the spilling of their blood and taking of their wealth, and the portrayal of them as being a small group outside the law and deviating from the path of the Believers whose ideas are ideas of extremism, militancy, isolation and harshness with no connection to the mercy of Islam and its tolerance and leniency.

And among the amusing things I have heard in this regard is what some of those who are called analysts and experts in Islamic groups said: that the constitution of al-Qaida Organization calls for the killing of anyone who breaks away from it!

We tell these slanderers who aren't ashamed to spread blatant lies, "Produce your proof, if you are truthful."

Show us this constitution in which you found this passage, and we guarantee to you that we will distribute it at the widest level and on all fronts. Al-Qaida Organization and its leaders are too noble and pure to descend to the rotten level of such nonsense.

They're not committers of random deed and impulsive acts or callers to imaginary temporal interests, nor do they tread aimlessly in their march; rather they base their methodology on clear legal evidence, well-founded Islamic fundamentals and definite doctrinal constants, and anyone can see the milestones of their journey, on one condition: that he be unbiased in his search for the truth and facts and not be blinded by hostility, jealousy or sheer ignorance.

Also from the methods which we indicated is concentrating on interpretive issues followed by the Mujahideen on the basis of legal interpretation and actual need and making them the central focus of criticism and considering them to be categorically flagrant mistakes which can't be accepted or rectified, and moreover, moving by way of these ["mistakes"] to pass unfair judgments without verification, deliberation or proof.

And the most prominent example of this is the bombings with which the Mujahideen target dens of infidel legislation, centers of criminal intelligence, military barracks and elsewhere, as occurred in Algeria and before that on the Arabian Peninsula.

These blessed acts are portrayed as having targeted first and foremost -

if not solely - the general public and the powerless, and the criminality and compound apostasy which was targeted hides behind a scene which moves emotions and whips up storms and is carried by the media.

Third: among the greatest methods used in the ideological war is the issuing of fatwas - or rather, the procuring of fatwas - which criminalize Jihad and the Mujahideen and describe them with well-known, repulsive legal terms like "bandits" and "Kharijites" and even "Qaramites," "extreme fanatics" and the like, and paint them with allegations of treachery and treason.

These muftis have become experts in perverting the source texts and are accustomed to bending their necks and don't even see anything wrong with occasionally breaking them if they refuse to be flexible.

In fact, states now form special committees of Shaykhs to debate the Mujahideen repressed behind the prison bars, like what is happening on the Arabian Peninusula, where they are following the tradition of the Egyptian government.

Tell me, what do you expect from someone who sees the sword above him, the rug in front of him and the Shaykh dictating to him the proof and evidence for the obligation of obeying the ruler?

For so long the Mujahideen and their 'Ulama have called for open, public debate without conditions or restrictions, so why didn't these 'Ulama accept the offer and confront evidence with evidence before the hands were placed in shackles?

And in a further attempt by the region's traitors to besiege the Jihadist methodology, there are rapid and continuous attempts to codify the sources of the fatwa and prohibit and criminalize deviation in the issuing or requesting of fatwas from the channels which they will specify. And the mission of those official channels will be to chant the praises of the tyrants, beat the drum for them and justify their hideous deeds while at the same time slandering the Mujahideen, stirring up misconceptions about their actions and issuing stringent fatwas against them.

Fourth, strengthening and backing some of the methodologies adopted by Islamic movements far removed from Jihad, especially those with a democratic approach and those groups which melt and bend the source texts and iron them out so that they agree with the civilization, culture and methodologies of the West, and portraying these groups as the moderate, balanced, reasonable and civilized alternative, and accordingly, pushing these groups into ideological conflict with the Jihadist groups, feeding that conflict and busying the Mujahideen with it.

This is one of the steps meant to isolate the Mujahideen inside their societies and place them in front of a torrential flood of ideas and methodologies which find backing, empowerment and publicity from numerous parties.

And finally, when those groups' mission is over, a cold shoulder will be turned to them, upon which they will realize they were only used to harm their brothers.

Fifth: killing, capturing, incapacitating or defaming the guiding Jihadi symbols, isolating them and preventing their voice from reaching the people, and emptying the arena of them or restricting them as much as possible, after which the Mujahideen will be without an authority in which they can put full confidence and which will direct and guide them, allay their misconceptions, and regulate their march with knowledge, understanding and wisdom, which in turn will lead to the intervention of some of those who have not fully matured on this path or are hostile to them in the first place, to spread whatever ideas and opinions they want to cause disarray and darkness in the proper outlook which every Mujahid must have.

Sixth: blowing out of proportion some of the minor, interpretive disputes which might occur among the Mujahideen, and considering them to be doctrinal/methodological disputes, and inventing new names and descriptions for those groups on the basis of these disputes, and making it an inroad for them to fan the flames of differences, bandy about allegations and spread rumors, to be able - by way of that - to turn minor interpretive disputes open to discussion into something portrayed as being deep, contradictory methodological differences on the basis of which groups are categorized as being moderate, middle-of-the-road or extreme.

And without a doubt, the atmosphere that reaches this degree of tension become a protected incubator and safe have for rumormongers, deserters and demoralizers, and he door is thrown wide open for defamation, casting of doubts, making of accusations and slander.

And at that point, however much the Mujahideen try to explain the truth, remove the misconception and reply to the accusations, their voice will be like the voice of someone with a hoarse throat in the middle of thousands of people shouting with one voice, a voice which is today represented by all the media without exception.

And Allah we ask for help.

And I say that what these enemies call the "war of ideas," and consider to be something new invented by their minds, is a type of war pursued by the minds of infidel Quraysh's champions and those hypocrite serpents who followed in their footsteps from the outset of the call of the Prophet (peace and blessings of Allah be upon him), and it is a war which we might call "the war of defamation, casting of doubts and distortion."

After they were beaten by argument and overcome by clear proof and were unable to confront it and their weakness in front of it became clear, they impugned its carrier and preacher, saying about him, "a sorcerer, a great liar," or "tutored, a man possessed," or "a soothsayer," as the Quran related from their predecessors: "In the same way there came no Messenger to those before them except that they said. 'A sorcerer or a madman." (51:52)

And when this argument didn't succeed and they found themselves the objects of derision due to their making of it - because everyone bore witness to the truthfulness of the Prophet (peace and blessings of Allah be upon him) and his trustworthiness, the integrity of his intellect, the clarity of his diction and the comprehensiveness of his call - they used the police of incitation, interference, shouting and abuse to prevent the people from listening to his voice and hearing his arguments, which is the argument of the bankrupt every time and is identical to what is perpetrated by the lowly media - whatever their inclinations and techniques.

"And those who disbelieve say, 'Listen not to this Quran, but make noise during its recital that you may have the upper hand.'" (41:26)

Interviewer: What do you mean by the methodology of melting and bending which you mentioned?

al-Libi: This methodology is one of the things with which our Islamic Ummah has been afflicted in this era and with which many of those who are described as thinkers, enlightened ones or moderates have been seduced.

The essence of this methodology is the melting of the facts of the religion and their bending and subdual to agree with personal understandings, outlooks and convictions, especially with regard to the issues which are felt to oppose or clash with some of the issues which the Westerners have publicized.

Thus, we find these benders sparing no effort and leaving no stone unturned to prove the agreement of Islam with what these Westerners believe and the identicalness of its rules to it and even its spreading of it. And regrettably, this virus has afflicted many of the Muslims' 'Ulama and callers, and they have become the heads of this school; the same 'Ulama upon whose shoulders Allah placed the preaching of the Message and its dissemination among the people as is without distortion or fraud and without them being deterred from that by temptation or intimidation, as Allah said, "Those who deliver the Messages of Allah and fear Him and fear none but Allah. And sufficient is Allah as a reckoner." (33:39)

So this school of melting and bending has turned all issues of religion into legitimate territory for their intellects to trespass on, their ideas to lick and their research to tackle, and no deterrent or barrier comes between them and that.

And this is a major disaster from our modern disasters, and the Quran has explained this truth and revealed the innermost thoughts of its people and their motivations for what they do, i.e. perversity and deviance of the heart, which we summarize here in being seduced by the culture of the West and being pulled by the temptation of thought and speculation.

Allah, the Glorious and Great, says, "As for those in whose hearts is perversity, they pursue that of it which is susceptible to different interpretations, seeking to cause discord and seeking wrong interpretations of it." (3:7)

The Messenger of Allah (peace and the blessing of Allah be upon him) recited this verse, then said, "If you should see those who pursue that of it which is susceptible to different interpretations, then those are the ones who Allah named, so beware of them."

Thus, the Muslim - in order to save his religion and protect himself from perversity and being lost in the wilderness of misguidance - must beware of them and warn against them and not be deceived by the names, titles, fame and so on, because what they are engaged in is the polluting of our Islamic sources and the desecration of our unique and pure concepts, which Islam has been keen to preserve with their independence in their terms, meanings and content, to prevent the leaking of any impurity to cloud this pure spring. And in this regard is the story of the Prophet (peace be upon him), when Umar (with whom Allah was pleased) came to him and said, "We hear from the Jews Ahadeeth which we like. Do you permit us to write down some of them?"

He replied, "Are all of you also confused like the Jews and Nazarenes were confused? I have brought it to you pure and white, and were Moses alive, he would have no choice but to follow it."

And today, you see that no sooner has a Western thinker or intellectual opened his mouth and praised something which Islam brought than it is turned into proof of the credibility of Islam, as if we are in need of these sorts of testimonies. And even worse and more dangerous than that is that these who have these loose ideas have started to bend Islamic concepts for these concepts to be a true witness to the correctness of many facets of the Western culture which clashes totally with Islam in meaning as well as in terminology.

I believe that we are in need of a sufficient, comprehensive study of the basics and foundations of this school of perversity and falsification in order to pull it up by the roots and make clear the danger it poses to the axioms of the religion, the extent of its corruption of Islamic concepts and its total opposition to the path of surrender of the heart and regulation of understanding taken by the predecessors and the well-versed 'Ulama after them.

I also bring to the attention of those who have shouldered the sin of strengthening the notions of this school and have spread corruption in the facts of the religion to curry favor with the infidel West, that this will not make them pleased with you, and that the day of your really big loss is when you hear the applause of the infidel West for your ideas and its satisfaction with your culture.

They want us to get closer to them by compromising on our religion one thing after another and untying its knots one by one, and they see nothing wrong with exuding for that - even if temporarily - some flexibility and open-heartedness, but without compromising in reality on any of their doctrines, constants and ideas, as Allah (the Glorious and Great) said, "They wish that you should compromise so that they may also compromise." (68:9)

And He (Exalted is He) said, "And neither the Jews nor the Christians will ever be pleased with you unless you follow their creed." (2:120)

So we tell these benders who melt the facts of the religion what our Lord (the Glorious and Great) said: "Say, 'Verily, Allah's guidance alone is the true guidance.' And if you follow their evil desires after the knowledge has come to you, you shall have [against] Allah no friend nor helper [to protect from Him]." (2:120)

Interviewer: You mentioned in the first pivot of the ideological war that Jihad is built on fixed bases and firm foundations, and that it is these which the Crusader enemy and his underlings are trying to undermine. Can you explain to us the most important of those fundamentals and pillars?

al-Libi: The first of these foundations and most important of them, in my view, is the definite belief that Jihad is a legal act of worship, and in fact, is one of the greatest acts of worship commanded by Allah in His Book and by His Messenger (peace and blessing of Allah be upon him) in his Sunnah.

Understanding this concept deeply and correctly automatically makes a person treat Jihad in the same way he treats prayer, fasting, the Hajj and other rites of Islam. Jihad isn't an option equivalent to other options to be picked from among them. And Jihad isn't an alternative to other independent means which have their own identity and boundaries.

Interviewer: Excuse me for interrupting, but isn't Jihad a means and not an end?

al-Libi: This phrase is correct if it is correctly understood.

However, many of those who repeat it error in appreciating its substance, perhaps intentionally. Jihad is a means to the greatest of ends and goals, which is unifying Allah (the Glorious and Great) and achieving His worship on earth, which will only completely and comprehensively happen through Jihad, and the nobility of the means is on proportion to the nobility of the end.

Allah (the Glorious and Great) said, "And fight them until there is no Fitnah [polytheism], and religion is wholly for Allah." (8:39): i.e. unless you fight them, there will be Fitnah, and Fitnah is infidelity and polytheism, as the interpreters have said.

And in the agreed-upon Hadeeth from the Ibn Umar (with whom Allah was pleased), he said that the Messenger of Allah (peace be upon him) said, "I have been ordered to fight the people until they testify that there is no God but Allah and that Muhammad is the Messenger of AllahÉ" to the end of the Hadeeth.

And Jihad s also a means to rescue the weak and oppressed and lift tyranny and repression form them, as Allah (the Glorious and Great) said, "And what is the mater with you, that you do not fight in the path of Allah and of the weak and oppressed men, women, and children whose cry is, 'Our Lord! Rescue us from this village whose people are oppressors, and raise for us from You one who will protect, and raise for us from You one who will help.'" (4:75)

However, despite Jihad being a means through which these noble ends and others are reached, Allah the Most High has made it an end in and of itself, when looking at it from another perspective: with it the people of faith are purified, the bad is separated from the good, the status of the truthful and sincere is raised and the lights of guidance and paths of success are opened for the believer in the affairs of his religion.

Allah (the Glorious and Great) says, "And We will, surely, try you, until We make manifest those among you who strive and those who are steadfast. And We will make known the true facts about you." (47:31)

And He said, "Do you suppose that you will enter Heaven while Allah ahs not yet caused to be distinguished those of you that strive and has not yet caused to be distinguished the steadfast." (3:142)

And He said, "And as for those who strive [and fight] for us, We will surely guide them to Our ways. And, verily, Allah is with those who do good." (29:69)

And since establishing the religion and empowering the Law and removing the humiliation which the Muslims are suffering is by no means whatsoever possible except through Jihad in the Path of Allah, and since it is the only way to achieve these ends, what is the point of the continuous droning about making Jihad a means and not a end, while using an erroneous definition of "means" in this context?

The Prophet (peace and blessings of Allah be upon him) said, "No people abandon Jihad without Allah covering them with torment."

And he (peace and blessings of Allah be upon him) said, "Whoever dies without raiding or hoping to raid dies on a branch of hypocrisy."

When the jurists (may Allah have mercy on them) divided acts of worship into "those intended in and of themselves" and "those intended for others," they never even imagined the definition of "means" which is held up by many contemporaries, who have made this phrase a cushion from which to abandon Jihad, shirk its burdens and search for alternatives with which they claim they will reach the same goal the Jihad will reach.

Regrettably, would that they had protected that goal form desecration and being disturbed by the minds and desires, but instead, they corrupted the end in the same way they perverted the means, and were negligent with the objective just like they took lightly what leads to it, and that is only because of their stripping Jihad of its devotional meaning and cutting it off from the world of the hereafter.

And the fact is, there is no act of worship which isn't an objective in and of itself from one perspective and a means from another perspective.

Prayer, for example, is a means to prevent indecency and evil, as Allah the Exalted says, "Éand observe Prayer. Surely, Prayer restrains one from indecency and manifest evil." (29:45)

But let no one say, "I've obtained my desired amount of refraining from indecency and evil in another way not involving prayer," and so abandon it because of that, because we say that prayer is an act of worship which Islam ordered, forbade its abandonment, warned against taking it lightly, threatened the one negligent in performing it, and laid down etiquette and rules concerning it.

And the same goes for every act of worship: it is a means to win the pleasure of Allah (the Glorious and Great) and win His Gardens and affection and be endowed with fear of Him, as Allah (the Glorious and Great) said, "O people! Worship your Lord Who created you and those before you, that you may guard [against the Fire]." (2:21)

And at the same time, it is meant in and of itself, whether as an obligation or a desirable deed, and Jihad is one of these acts of worship, and in fact, is at their peak and at the top of their pyramid, as he (peace and blessings of Allah be upon him) said, "And the pinnacle of its hump is Jihad."

Interviewer: Let us return to the issue of the pillars on which the Jihadi methodology stands.

al-Libi: Yes. I said that the first foundation and main base in this issue is complete faith and definite belief that Jihad is a divine devotion to which we have been commanded and that we must make every effort to discharge it. in the same way we discharge prayer, fasting, Haj and other devotions.

And many of the groups which took the path of Jihad first were beset by blemishes and flaws because of their confused understanding of this fact, as Jihad was - or became - in their imagination a mere dry means, like other earthy means by way of which strive to establish states.

So when darkness permeated this concept, it was easy for them to abandon Jihad and jump to the ballot boxes as another option equal to the devotion of Jihad, and therefore an alternative to it.

Thus, we see in contemporary terminology the popularization of terms like "the option of Jihad," "the option of resistance" (by which they also mean Jihad), "the option of arms," "the option of combat," "the option of struggle," and so on to the last of this disgusting modern series.

As for us, we say "the devotion of Jihad," and when we add "devotion" to Jihad, this makes it spring from submission, obedience, surrender and abandonment of choice when ordered.

As for the second pillar on which the Jihadist methodology stands, It is that this devotion which we perform and which we make every effort we can to bring to life has as its first objective the establishment of the religion, rule of the Law and making the Creation worship their Creator.

So it is a noble means to a great end which is not dirtied by the pollution of patriotism, the filth of nationalism, the stain of rationalism or the rigging of whims.

And it is the lofty concept which the Quran and Sunnah express with the phrase "in the path of Allah": i.e. in obedience to Allah (the Glorious and Great).

And this motive must be clearly defined in the depths of the soul, just as it must be active and tangible in practical pursuits.

This is why it is related in the authentic Hadeeth from Abu Musa that he said: "A man came to the Prophet (peace be upon him) and said, 'One man fights for booty, another man fights for fame, and another fights to show off, so which of them is in Allah's path?' He replied, 'The one who fights for the supremacy of Allah's Word is in Allah's path.'" And he (peace and blessings of Allah be upon him) said, "He who fights under an ambiguous banner out of partisan anger or calls to partisanship or aids partisanship and is killed, is one killed in the Time of Ignorance."

And on the basis of this fundamental, we caution some of the Islamic groups, among them HAMAS, which are risking the blood of their members and pushing them into battles which have reins or bridles, and we call on them to cleanse and empty their Jihad of contemporary Jahili pollutants and of the terms which Satan made to appear pretty to them but which are like wind in the desert on the balance of Law: e.g. patriotism, nationalism, shared destiny, the supreme interest and the other slogans which are repeated many times daily on the tongues of their officials and commanders.

None of this has any place in the religion of Allah (the Glorious and Great), and it is one of the major catastrophes resulting from the failure to appreciate this genuine pillar of the Jihadist methodology.

Therefore, let our intent be clear, our target defined and our objective declared: the establishment of Allah's religion in its complete and comprehensive sense which is articulated by the statement of Allah, "and religion is wholly for Allah." And it is an objective for which we make sacrifices and for whose establishment we pour out our blood and give of our efforts, and everything else, whatever its value, is less than it.

And that's why we don't accept the least compromise regarding it, nor do we agree to make it the subject of research, review and giving and taking, even if holding on to it leads to the total annihilation of our groups.

We aren't better in that regard than the People of the Ditch.

So there is no room for respecting international legality, nor for observing the charters of the United Nations, nor for referring to the resolutions of the Security Council, nor for sanctifying the charter of the League of Arab States.

All of these organizations with everything and everyone in them aren't worth an ant to us.

As for the third pillar on which the Jihadist methodology stands, it is loyalty and disloyalty, in doctrine and concept and behavior and action.

This deep Imani concept which is the strongest knot of Iman, as the authentic Hadeeth says, is among the most important and greatest of the things on which the Jihadist methodology is founded. And any scratching or playing with it means to shake and disturb the Jihadi March from the inside and liquefy it in an ugly way by giving precedence to imaginary so-called interests with which the fundamental of the religion and its mainstay is destroyed.

Loyalty, from which branches out the right of help, aid, love and affection, must be based on one foundation as firm as the mountains: Iman.

The Muslims are one Ummah, from their east to their west and from their north to their south, and equal in that are the red and the black, the Arab and the non-Arab and the near and the far.

Allah (the Glorious and Great) said, "And the believers, men and women, are protectors, one of another." (9:71)

And he (peace and blessings of Allah be upon him) said, "The parable of the believers in their mutual love, mercy and sympathy is that of the body: if one of its limbs falls ill, the rest of the body responds to it with sleeplessness and fever."

So there is no consideration given to color, not to race, nor to tribe, nor to borders, nor to farness or nearness, and it is the same whether he is in Africa or Asia or Europe or America or Australia. Rather, [the differentiator] is faith and piety, and the one who is pulled back by his deeds is not pushed forward by his affiliation.

And the rooting of this concept in our hearts obligates us to adopt the causes of the Muslims and have the real feeling which leads to action that their happiness is our happiness, their sadness our sadness, their tragedy our tragedy and their security our security.

And there is nothing connected with the Muslims which we could possibly call an internal issue and thus shirk our duty of helping when we are able, declaring our loyalty and discharging the rights of faith-based brotherhood.

So the general rule which stems from this deep-rooted concept is what the Prophet (peace and blessing of Allah be upon him) made clear with his statement, "Help your brother, be he the oppressor or the oppressed."

A man said, "O Messenger of Allah, I can help him when oppressed, but how can I help him when he is the oppressor?"

He replied, "Prevent him from oppressing, and by that you have helped him."

And the Hadeeth is agreed upon.

As for the concept of disloyalty, which implies unending hostility and continuous hatred between faith and its people on one side and infidelity and its party on the other, then the infidels - all the infidels - are our enemies, and we neither are loyal to them nor befriend them nor love them.

How can we love or be friendly with those whom Allah has made His enemy?

No, we treat them as the Imam of the Haneefs and father of the Prophets, Abraham (peace be upon him) said, "We have nothing to do with you and with that which you worship beside Allah. We have rejected you [i.e. your religion and way], and there has begun between us and you enmity and hatred forever, unless you believe in Allah Alone." (60:4)

So we don't distinguish between a patriotic infidel and an alien infidel, nor between a local infidel and foreign infidel, nor between an invading occupier and an apostate helper.

Accordingly, we are not of those who make the ties of allegiance nationality, citizenship, relationship or so on.

The believer is our brother, even if he be the farthest of the far in his dwelling and affiliation, while the infidel is our enemy, even if he be the nearest of the near in his district and connections.

Allah (the Glorious and Great) says, "O you who believe! Take not for allies your fathers and your brothers if they prefer infidelity to faith; and whoever of you take them as allies, they are the wrongdoers." (9:23)

And He (the Exalted) says, "You shall not find any people who believe in Allah and the Last Day loving those who oppose Allah and His Messenger, even if they be their fathers, or their sons, or their brothers, or their clan." (58:22)

And He (the Glorious and Great) says, "Your allies are Allah, His Messenger, and the believers, those who establish regular prayers and pay Zakat and bow [in worship]." (5:55)

We are also not those who divide the Jihad and allow, support and call to it against the Jews in Palestine but forbid, criminalize and prevent and deter it in Iraq or Afghanistan or Chechnya or Algeria or elsewhere. Jihad, which is the highest and the greatest form of disassociation, is against the Jews in the exact same way it is against the Nazarenes, Magians, Hindus and apostates, because the type of relationship we have with all of these sects is one: total separation, complete disassociation and open animosity.

And as long as the matter is thus, we will not prefer one infidel to another because of his patriotism or nationalism or his noble descent or affiliation.

We fight the polytheists one and all as they fight us one and all, and we don't stop in that at any boundary, or limit ourselves to one type, or confine ourselves to any territory.

This is our path until everyone submits to the religion of Allah (the Glorious and Great) and complies with its rules and surrenders to its authority.

Interviewer: Does this mean that you will open battlefronts with all of these sects which you mentioned at the same time?

al-Libi: No, that's not what I mean, nor is it legally or rationally desirable. I haven't been talking about the issue of fighting from the perspective of when we should begin, and with whom we should begin and how we should begin.

That is a legal question subject to legal policy, based on the predominant interest and dependant on ability.

And it is a question of arranging priorities which Islamic Shari'ah has made clear, as Allah (the Glorious and the Great) has said, "O you who believe! Fight the unbelievers who are near to you and let them find harshness in you; and know that Allah is with those who fear Him." (9:123)

Jihad, like other acts of worship, is contingent on capacity and capability, as "Allah burdens not any soul beyond its capacity." Rather, my talk dealt with the nature of the relationship on which Islam is founded, between its people belonging to it and between those other than them from the people of the other religions, that relationship which, whenever it is on the correct, clear doctrinal basis, the Jihadi march is disciplined and free of the methodology of melting and bending, which now has its callers, thinkers and theorists.

Yes, we believe that all of earth must be under the rule of Islam, with no exception made for the smallest part of it, because our Messenger (peace and blessings of Allah be upon him) was sent to all the people without exception.

But this in no way means that we will fight all of the peoples of the earth at one go to subjugate them to Islamic Shari'ah. Islam didn't order us to do that, but rather, ordered us to fight the nearest, then the next nearest of those who refuse to submit to Islamic rule and begin with the closest, then the next closest.

And in this way, the circle widens until all submit to the rule of Allah.

And we are now at the first step and beginning of the road, as we are striving to recover our territories taken over by the infidels: the Jews and Nazarenes and their apostate helpers, the traitorous rulers.

And this is the duty of the Muslims today, to find for themselves a foothold where they can establish their state which will rule them by Islam and under which they will be shaded and in whose justice they will bask.

Interviewer: Speaking of the issue of priorities in fighting, there are some who propose beginning with fighting the apostate governments, due to them being the enemy closest to the Muslims, instead of the Americans and the other infidel coalitions.

al-Libi: Without a doubt, the original confirmed ruling laid down by the noble verse and attested to by the biography of the Prophet (peace and blessings of Allah be upon him) and practiced by his Companions after him is that we begin fighting the nearest, then the next nearest, as Allah (the Glorious and Great) said, "O you who believe! Fight the unbelievers who are near to you." (9:123)

But this is when the situation is uniform and regular, i.e. when things are going normally, so that the Mujahideen move in their conquests from the nearest to those adjoining them. And this is one of the strongest proofs that Jihad is not stopped by borders nor limited to the defensive form alone, as many modern defeatists try to establish.

In any case, the jurists who spoke about this issue and clarified its ruling stipulated that there are numerous situations in which it is better to start by fighting the farther enemy and giving priority to him over the others.

And among those situations is if the farther enemy is more harmful and dangerous to the Muslims and their religion, and assessing this is the job of the commanders of the Mujahideen, who decide - after consultation and review - which of the enemies were more deserving of beginning with, in light of any of the accepted legal considerations.

So the issue isn't a definite textual issue not open to discretion, study and preference according to reality, need, ability and interest. Rather, Law has given a free hand in it and given consideration to interest and assessment.

In addition, the question of special nearness and farness in our modern era doesn't have the same significance it once had when we look at the facts, because the type of weapons used - aircraft, missiles and so on - have penetrated borders and broken through barriers and now cross continents and oceans and target the Muslims as they sit in their houses with their families.

And the relations which tie the major infidel states to the statelets and their apostate governments are close, overlapping relations on all fronts: political, economic, military and even cultural.

So in general, they are a single entity, a single enemy and a single army, and they area single hand against us and the battle they are waging against us is a single battle which either the infidel Crusader states adopt themselves or is taken up by their traitorous proxies who reign over the Muslim peoples.

This is in addition to the heavy military presence of those states - foremost among them America - on the soil of the Muslims, killing sabotaging and destroying, and violating their sanctities, plundering their treasures and imposing on them their policies and laws.

And these apostate regimes with these states are like troops with their commander, or rather, like slaves with their master: making not a sound and speaking not a whisper.

And all of us know that smashing this modern idol and inflicting defeat on it automatically means the weakening of these emaciated regimes of treason, which will be buried with their god to whom they were devoted and be thrown with him into the trash heap of history, without being mourned.

So the Mujahideen today are in the situation of repelling the enemy and stopping his fierce attack on the Muslim lands, and thus the option of beginning fighting with this enemy or that doesn't really have much meaning now. And even the one who wants to begin by fighting the apostate regimes dominating the Muslim lands will find himself after a little while - if not from day one - confronting in one way or another the Crusader forces, foremost among them America. And thus he will stand face-to-face with the enemy he used to consider and suppose to be the farther and avoided fighting first.

So with our enemies today, their near is near and their far is near, and the part the Mujahideen play in choosing the time of the confrontation is to try and aim as much as they can to enter the decisive battle which suits their abilities, has the factors for success, saves them from a lot of effort and leads to the elimination of the greater enemy which spreads corruption ad ruin and under whose wings the regimes of tyranny and torture develop and prosper.

Interviewer: You indicated previously that it is obligatory on the Muslims in general and the Mujahideen in particular to bring into being a place or state which will be their primary base for spreading the religion of Islam all over the world. And as you know, the Mujahideen in Iraq have announces the setting up of the Islamic State of Iraq.

So what is your perspective on this step taken by our brothers there?

al-Libi: The fact is, I consider the venturing of our brothers, the Mujahideen in Iraq, to declare the setting up of the Islamic State to be pure God-given success and to be part of the guidance which Allah has guaranteed for His worshippers the Mujahideen in His statement, "And as for those who strive [and fight] for us, We will surely guide them to Our ways." (29:69)

In fact, I have no doubt that it is from Allah's defense of the believers who help His religion and Book, as He (the Glorious and Great) says, "Verily, Allah defends those who believe. Verily, Allah does not love every unfaithful, ungrateful one." (22:38)

So the Jihad in Iraq before the declaration of the State was moving toward a dangerous, deadly slide, but stealthily and secretly. And what revealed this slide and uncovered that lethal abyss was the announcement of the setting up of the State.

And as a result, the enemy occupier fell into an unenviable dilemma, as this step upset all his calculations and shuffled all his cards, because the plan which the enemy was going on before the declaration of the State was the creation of a puppet government of Sunnis which would be somewhat acceptable and recognized by the neighboring weevil states, especially the stats of Al Saud and Jordan, both of which played the largest part in this conspiracy and began to blow the trumpet of defending the Sunnis in Iraq and pretended to be concerned for their lives, and the media began to enlarge this issue and produce it in various forms.

And we don't mean that our brothers, the people of the Sunnah in Iraq, weren't - and aren't - suffering the ugliest forms of both Rejectionist and Crusader criminality.

No, what I mean is that the constant droning about this issue by states audacious in their hypocrisy and criminality was part of a major conspiracy being hatched against the Jihad and Mujahideen in Iraq.

After the torture and killing of the people of the Sunnah had reached its utmost degree and everyone in Iraq and outside it had been convinced that they had reached the pinnacle of suffering, a political drama would be fabricated for them to produce a government with a Sunni majority, and the states of the region would rush to support and strengthen it and polish its picture, for it to be said to the people of the Sunnah in Iraq, "Here, you've got what you wanted and won what you were asking for, and you've been saved from the abattoirs of the Rejectionists and the massacres of the Crusaders, so enjoy a secular puppet government for which near and far have beaten the drum."

And whoever would object to that so-called "Sunni" government would find himself an eccentric outcast, as he would be trying with his actions to destroy this big palace won by the people of the Sunnah in Iraq and under whose umbrella and protection they live.

And in this way the curtain would fall on the sacrifices of the champions, tears of the windows and suffering of the orphans, to be blown away by the storm of celebration, support, polishing and artificial heroics with which that hypothetical government would appear.

But Allah saved and rescued the Iraqi Jihad from a crushing blow which would have brought it back to square one, with the declaration of the setting up of the Islamic State of Iraq.

Interviewer: But as you know, many of the Mujahideen inside and outside of Iraq believe that this step wasn't in the interest of the Jihad and Mujahideen.

al-Libi: As far as I can see - and Allah knows best - the objection by these virtuous ones to the declaration of the Islamic State of Iraq had been blown out of proportion. What our brothers embarked on in terms of the declaration is a step the least that could be said about it is that it is the discretionary judgment of a large group and wide slice of the Mujahideen, and it should be dealt with on this basis and we should move on from this stage to the stage of direction, guidance, backing, strengthening, closing of ranks and constant striving for what is better and fuller, instead of us stopping at the point of the declaration and portraying it as if it is the deathblow for the Jihad in Iraq, even though the facts on the ground demonstrated otherwise and reveal the major positive aspects which have appeared one after the other after the declaration of the state.

And we shouldn't close our eyes to the biggest gain of this declaration, which is the rescuing of the Jihad in Iraq from an eradication program which would have destroyed its foundations, and this is the fact which the enemy appreciated before the friend.

So to say that this step was not in the interest of the Jihad and Mujahideen isn't correct at all, and is a feigning of ignorance which clashes directly with the reality in which the caravan of Jihad in Iraq is moving. And the fact is, when we want to assess anything and arrive at a correct and fair result, we must weigh the positive and negative aspects which this act or that includes.

"They ask you concerning wine and games of chance. Say, 'In both there is great sin and harm and also some advantages for men, but their sin and harm are greater than their advantage." (2:219)

So it may be that one positive aspect, due to its weight, strength and predominance, downs hundreds of negative aspects in its wake, and vice versa.

And those who talk about the great project of Islamic civilization and cling to the model form of it and make it their gauge in the success or failure of any work have not really gone down to the realistic field of work, and haven't come into contact with the details of events, and have not looked deeply at the degree of plotting, deception ad intrigues which the minds of the chiefs of infidelity produce day after day.

Tell me, for the sake of Allah, will this project of civilization which the minds enjoy imagining and swim in the seas of its fantasy be born overnight, complete and strong and filling the horizon with its prosperity, civilization and vigor? Or will the goings-on of reality and the directions of events be totally reflected on it in its strength, size and expanse, except that it will resist and be resisted, will do its best and will get stronger and ascend until it approaches slowly but surely the stage of perfection, which will only be after a period of decades and not in a day or two?

Interviewer: In your view, what is the most important of the dangers which the brothers in the Islamic State of Iraq face at this stage?

al-Libi: As I told you, the announcement of the State was a huge surprise to the enemy occupier by all standards, as the automatic conviction of the Iraqi people became that the State is the immediate and only replacement which will succeed the occupier when he pulls out, Allah permitting.

And with the consecutive blows taken by the Crusaders at the hands of the Mujahideen there and the increasing internal pressure for the Bush administration, and with the severe threat they feel from the Muslims' setting up a state with full independence at all levels, the concern of the occupier has become making this project fail and burying it alive in the cradle in any way possible, including pulling out this conviction which has settled in the hearts of the Iraqis with the declaration of the State, to create for them after that any replacements they wish.

And the most successful method in this regard was spreading and feeding internal differences among the Mujahideen; exploiting some of the points on which their outlooks differ, enlarging them and deepening the rift through them; attempting to infiltrate the ranks to play with the Jihadist methodology from the inside; and committing some ugly actions and attributing them to the Mujahideen to repel people from them. And with that, the Mujahideen will become preoccupied by themselves and busy with the endless chain of their own problems, and their energies will be depleted and their efforts wasted and the trust between them will disappear.

Tis is why I liken the situation of our brothers in the Islamic State of Iraq to that of someone walking in a minefield in pitch-black darkness, and this requires of them total alertness, constant probing of the danger spots, comprehensive awareness of the types of conspiracies and insight which reveals to them the way of the criminals. And it requires that their decisions spring from a comprehensive view of events, independent treatment of issues, sensing of responsibility at every step and total aloofness from the policy of reactions and responding to provocation, because this is an extremely critical and delicate stage, and emerging from it safely means full victory and total empowerment, with the permission of Allah, the Most High.

Interviewer: The fact is, there are many hot issues which we wanted to take up in this interview, like the issues of Palestine, Algeria, Somalia, the Arabian Peninsula and others. And perhaps another opportunity will be had for the talk to be more comprehensive and complete.

But do you have a closing word to direct to the Mujahideen in particular and the Ummah in general?

al-Libi: I say that the stage which the global Jihad is passing through is one of the most critical stages, and the methods of confrontation used by the two camps aren't the least bit equal, but what worked to our advantage - after taking the accepted precautions - is our honesty with Allah (the Glorious and Great), and our deep and unqualified faith that Allah is with us, and this requires from us submission to Allah and humbleness in front of Him and distancing ourselves from arrogance, vanity and conceit.

This is the most important of the things with which we advise our brothers the Mujahideen in all arenas of Jihad.

And their trust in Allah should be great and their conviction of the correctness of their methodology firm, and they should put in front of their eyes the statement of Allah (the Most High) in which He reminds the noble Companions of His favoring them by aiding them on the Day of Badr despite their fewness and weakness: "And Allah had already helped you at Badr when you were few. SO fear Allah so that you may be grateful." (3:123)

And they of all people should be the most fearful of disobeying Allah, because is it logical for us to disobey our Lord (Exalted is He), then ask Him to help us?!

I also advise them to close their ranks and unite and bulldoze every obstacle which comes between them and that, because all dispute is evil, even if we imagined that there is benefit in it and we claimed through it "reform."

This is with preservation of the safety of the methodology and clarity of the objectives and cleansing them from any stain which sullies their purity, like Jahili fanaticism, nationalism, patriotism and the like.

If the infidels, who have contradictory religions and inharmonious hearts, have come together and agreed and compromised with each other and formed coalitions in consideration of their supreme interest represented by the elimination of Islam, then how about us, the Muslims, when our religion is one, our creed is one, our Lord is One, and our Prophet (peace and blessings of Allah be upon him) is one: aren't we more deserving of this honor and this virtue?

As for the Islamic Ummah, I tell it: rejoice, for by Allah, the darkness of tyranny and repression is about to fall away, and the suns of truth have begin to send their rays slowly but surely. And that wouldn't have happened with servility, submission and surrender and by fleeing from the places of danger and risk; no, this cloud began to lift through the sacrifices of your righteous sons, sacrifices which mixed the blood of the martyr with the sweat of the weary and the tears of the bereaved. SO have the patience, our beloved Ummah, for victory is an hour of patience, "and on that day the believers will rejoice in Allah's help. He helps whom He pleases, and He is the Mighty, the Merciful." (30:4-5)

Interviewer: As we close this interview, we thank Shaykh Abu Yahya, and we ask Allah to benefit through his knowledge and reward him well of behalf of Islam and Muslims.

al-Libi: And you too: may Allah reward you for your effort and striving, and we ask Allah to bless it.

And peace be upon you and the mercy of Allah and His blessings.

The Closing Statement for the Religious Training that was Held at One of the Mujahideen Centers

Release Date: 7 Nov. 2007
Production Date: Unknown
Type: Video statement
Version: Translation

Thanks be to God, that by His blessings and good deeds are complete, and "Praise be to Allah, who hath guided us to this (felicity): never could we have found guidance, had it not been for the guidance of Allah" [Part of a Koranic verse; al-A'raf 7:43].

Today is the 12th of the month of Jumada al-Thani of the year 1428 from the migration of the chosen [the prophet], peace and prayers be upon him. This day is the last day of the Religious Training Course that, we pray to God, Exalted He be, to save the days of the course for us to benefit for it in a day that no money or children will do us good but only those who come to God with pure hearts.

As I told you at the beginning of the Training Course, courses like this are an example for the human's life that has a beginning and an end. The ending is according to one's effort during his life. The training course as we saw acquired from us effort, study, and revision so the person obtains from it what he wants of knowledge that he came to obtain from the beginning.

By the grace of God, the Great and Almighty, upon us He combined for us between jihad worship and acquiring knowledge worship. This is from the precious blessings of God, the Great and Almighty, and He is the Generous, the Benefactor, and the Magnanimous that God has combined, Exalted He be, for His servant between the most honorable and great worship. For the jihad as we know is the greatest of worships as mentioned in the Hadith by the prophet, peace and prayers be upon him, as well as mentioned in the book of God, the Great and Almighty as follows: "Not equal are those believers who sit (at home) and receive no hurt, and those who strive and fight in the cause of Allah with their goods and their persons" [Koranic verse; al-Nisa 4:95] to the end of the verse. So one should include in his jihad the dedication to seek knowledge even for a few days and he should know that this is from God's blessings, Exalted He be, upon him. Why is that? It is because initially we were created to worship God Almighty.

The Almighty says: "I have only created Jinns and men, that they may serve Me" [Koranic verse; al-Zariyat 51:56]. The worship is an obligation from God, the Great and Almighty, meaning it is orders and forbiddance and there is no worship from all the worships whether is was prayers or fasting or Hajj or jihad that must have its own rules that the Muslim need to know them. So if God facilitates for you the doors to seek knowledge, then know that He facilitated for you to worship Him and you are acumen. The human has to worship God so if he does not worship God based on knowledge, understanding, insight, and proof then he will defiantly worship God on the bases of ignorance and blindness. I seek refuge in God.

So this is one of God's blessings upon us that we need to be aware of all the time. Then we make efforts to be thankful for them so we do not lose them. We know the Hadith of the prophet, peace and prayers be upon him, "There are two blessings in which most people are in great loss: 1. good health; and 2. free time" [Hadith]. You stayed at this house or this center all this time far from your family and far from your brothers dedicated for one thing only that is seeking knowledge. This is one of the greatest blessings that God, Exalted He be, has facilitated this for you.

Therefore, acquiring the knowledge of worship, which without a doubt is worship, then the worship of seeking knowledge and the worship of jihad is the greatest that is facilitated by God, Exalted He be, for His believer worshiper is the first priority.

What we extract from this training course is that the separation that some people are attempting to make between the worship of jihad and the worship on seeking knowledge has no base of truth, meaning the assumption of an existing jihad without knowledge is not true. And the assumption that the scholar and the student of knowledge cannot participate in jihad is not correct. Why? Because God Almighty informed us of the truth of His book and said "Had it been from other than God, they would surely have found therein much discrepancy" [Koranic verse; al-Nisa 4:82]. These are the verses of God Almighty which are in the Koran. As the book of God Almighty, which came down from Him, does not contain any contradiction within its verses and there are no differences within its rules, likewise the devotions which God Almighty ordered us to conduct do not contain any contradictions. There are no contradictions within them. So this matter that some tried to cultivate, invent, and give more importance to than necessary, that jihad and knowledge do not meet, is incorrect. The example or the proof of this is what we are currently going through. That with the grace of God Almighty is proof that the pursuit of knowledge is possible in the arena of jihad. How many of the jurisprudential rules that have no connection to jihad were handed down in the arenas of jihad? Of those are the verses of Tayammum. Where was the verse of Tayammum handed down? It was handed down when the prophet, prayers and peace be upon him, was on his way while returning from one of the raids. Isn't that right? Moreover, God Almighty informed us of this and said "Nor should the believers all go forth together if a contingent from every expedition remained behind, they could devote themselves to studies in religion, and admonish the people when they return to them" [Koranic verse; al-Tawbah 9:122].

Then God made it accessible for us by combining the jihad and the pursuit of knowledge and this is one of the greatest blessings with which God Almighty has graced us. This is the first matter of which you must be aware. The other thing, as we have mentioned, is that there is no contradiction between the pursuit of knowledge and jihad for the cause of God. We, as mujahideen, are required to learn the rules of our religion unlike other Muslims, and I do not mean only rules of our religion that pertain to worship through jihad. No. We are required to know the rules of prayer, the rules of purity, and the rules of fasting. It is imperative that the mujahid should know the rules regarding any devotion that God Almighty has decreed upon the believers and made a requirement. The mujahid is not a person who is exempted from the rules of God Almighty. He is not a person who is excused from legitimate religious obligations; rather, he is like all other Muslims, like all other Muslims.

As we are in the arenas of jihad, we are learning the rules of jihad, its behaviors, regulations, and its criteria. We are also required to learn the rules of the other forms of worship. What is the benefit of the jihad of a mujahid, and that he battles for the cause of God, when he does not know how to perform wudu? Or when he does not know how to pray? Or when he does not know how to fast? What is benefit of this? Does this mean that he worships God through his jihad clearly and knowingly while at the same time worshiping God in darkness, blindness, and fog in prayers and other matters? This cannot be possible. It cannot be possible. It is with the grace of God Almighty that he made us successful in proving truthfully and practically that these are not contradictory. Additionally, the book of God does not have any contradictions in its verses and no disjoints between its rules. Similarly the devotions that God Almighty ordered of us do not contradict or oppose each other. Is that clear? Good.

The third matter which I would like to attend to – and open your ears to this – is that the person who has taken this course and learned what God Almighty has handed down to him does not mean that he has graduated and become a person versed in jurisprudence or a mufti or a person who heads religious meetings handing down fatwas of "this is right" and "this is not right" or "this is haram" or "this is halal". No! We have only met here so that we may learn the rules of God to be able to worship God knowingly. As for the status of fatwa and the status that "this is halal" and "this is haram," this requires a long period of time for the human being to attain. This cannot be accomplished in a month or two months or even in a year or two years. Imam Malik – and he is who he is – says, "I did not sit in this gathering, meaning so that he may make fatwas, until I was confirmed by 80 scholars that I am able to make fatwas." So 80 scholars confirmed Imam Malik and told him: "You are now able to make fatwas." So now, who will confirm that we are able to make fatwas? And in a month or two months or a year or two years one of us may not even be well versed in the Fatihah or be able to distinguish between duties, recommended actions, or what is most hated or what is prohibited. You must fear God Almighty. I do not want to hear of one of you who left here to head up gatherings and now says "this is halal" and "this is haram." Yes, you are allowed to say "that which I have studied is such," but to make fatwas in verses that are handed down or in core matters or someone approaches you and asks your opinion and you tell him "this is permissible" or "this is prohibited" or "this is hateful," then that is not acceptable. [It is] not acceptable religiously and we also do not want it in our behavior or our manners. Be aware of this matter. This is one of the most important matters of which I want you to be careful. Have you understood this well? Have you understood this well? Yes.

Then these are the matters that I wanted to bring to your attention and what we gain from these courses, even if they are short, is to know the value of scholars, their status, and their eminence. Scholars are like the sun during midday. If they are present, people are able to live a good life and are able to conduct their lives in a normal manner. But if the sun sets, what will happen? What will befall the people? Darkness, it will be darkness.

Darkness is ignorance, blindness, temptation, and heresies, and thus they all are going astray. We realize the value of the ulema through courses like this one. Because you worked hard during a period of one month in memorization, repetition, revision, and studying, you feel that you worked hard to learn just a tiny, insignificant amount of knowledge. Yet you are rejoicing in what God, the Most High, has given you. Imagine how it is with a scholar who spends his entire life in seeking knowledge, teaching people, enduring revision, fatwas, and supplications to the Almighty, exalted He be? Where are you compared to that? We do not want to turn our gatherings into a place where we speak about the wicked, or agree with the ulema, but rather we want to know the value of these righteous ulema who fear God, convey His message, and fear no one but God. You see? It is imperative that you pay attention to this. If you see a good scholar who fears God, exalted He be, you must appreciate him and you must remember that you endured one month or two in seeking knowledge and found that it requires patience and revision in order not to lose the acquired knowledge. Imagine how it is for these scholars who might have spent their entire life doing the same. This makes you appreciate people and know their ranks; it makes you realize how much esteem and respect they deserve. As we have mentioned, God, exalted He be, made them his partners in witnessing to His Oneness, which is the greatest witness ever. The greatest witness indeed, "That is the witness of Allah, His angels, and those endued with knowledge, standing firm on justice. There is no god but He, the Exalted in Power, the Wise" [Koranic verse; al-Imran 3:18]. Do you understand?

The status of the ulema is great, and if they slip or err, God, the Most High, did not create His human creation free from fault. Who said that man does not err? Who said that man does not, sometimes, face temptations that sway him right and left. But in the religion Islam, our rules state that good deeds cancel bad deeds and not bad deeds cancel good deeds. You do not come to a scholar who spent his entire life and exerted all his energy in Da'wah to the Almighty and in teaching and educating the people in religion and then with one or two mistakes, you turn around and curse him and erase all his good deeds. This is not the scale of Sharia. It is the scale of whim, and we are not the people who follow our whims. Fairness is closer to piety. Do you understand? This is the third issue.

The fourth and last point is that God, exalted He be, has opened the door of the pursuit for knowledge to you, and you have discovered the joy of the search for knowledge and the beauty of patience in your perseverance and pursuit for knowledge. Since God has opened the door for you, rely on Him and continue in the path, exerting all that you can. Instead of spending part of your day or your life listening to [religious] songs, which do not feed or satisfy a hunger, spend the time in listening to lectures, studying, listening to a Fatwa tape, or studying with one of your classmates. Should there be another course, God willing, try your best to join it because knowledge comes one step at a time, and not in one installment.

The last point is to do your best in offering supplications to God, exalted He be, to bless you in what He has given you. Remember that knowledge is not measured by the amount one memorizes. There are many whom you might find like an encyclopedia or a computer and who can answer you in any subject you bring up, but God has not blessed their work; while you might find a person of modest and limited knowledge, who has combined his knowledge with sincerity and supplications as commanded [sound cuts off and then he continues] to ask the Almighty to grant him more knowledge.

I ask God, the Exalted and the Almighty, to make us among those who obey and follow the best of what they hear. We ask the Almighty to make us among those who benefit from His knowledge and we seek refuge in Him from the luring words and actions. He is the All-Hearing. Now, let us start, God willing, with revising the Hadiths. We can start from the right. After that, we will listen to a recitation from the book of God, exalted He be.

Going Forth

Release Date: 18 Dec. 2007
Production Date: Unknown
Type: Video statement
Version: Translation

Praise belongs to God. We praise Him and we thank Him. We praise Him and ask for His help and His forgiveness.

All Praise is due to Allah. We worship Him, We seek His help and His forgiveness and we turn to Him in repentance. And we seek refuge in Allah from the evil within ourselves and our ill deeds. Whoever Allah guides cannot be lead astray and whomever He leads astray, for him there is no guide. I testify that there is no god but God, with no partner, and that Mohammed is His servant, His messenger and the chosen from among His creation. "It is He Who hath sent His Messenger with guidance and the Religion of Truth, to proclaim it over all religion, even though the Pagans may detest (it)" [Koranic verse; al-Tawbah 9:33]. May the prayers of God be upon him and upon all his family and companions, and anyone who walked in his path until Judgment Day.

Thereafter, praise be to God, who guided us to this, without Whose guidance we would not have found our way. Praise be to God, who bestowed on us the blessing of Islam, the Koran and the blessing of following the master of the messengers, Mohammed, prayers and peace be upon him. Praise be to God, who blessed us with faith in Him, and favored us with emigration for His cause and with jihad. He, who decreed it in His book, and commanded it to His prophet, prayers and peace be upon him. These three matters – Islam, emigration and jihad – are the best blessings which God, exalted He be, bestowed on His servants. The prophet, prayers and peace be upon him, said, "Satan sits in the path (of every goodness) that the son of Adam (may try to take). He sat in front of him when he took the path to Islam and told him, 'how could you leave the religion of your fathers and your fore-fathers?' But he disobeyed him and became Muslim. Then Satan sat in front of him in his path to Hijrah and said, 'How could you leave your land and your sky?' But he disobeyed him and migrated (from Mecca). Then he sat in front of him in his path of jihad and said, 'Why should you do jihad? It will only exhaust your wealth and body. You'll be killed, your wife will marry someone else, and your wealth will be divided (to others).' But he disobeyed him and went for jihad. Rasul Allah then said, 'Whoever does this, it is a duty upon Allah that He shall enter him into Paradise' [Hadith]."

This great hadith shows us the correct path, which the servant takes in this world, by taking the right path and following the Sunnah and the guidance of the prophets and the messengers, first by entering in the religion of Islam, even though entering this religion and following its true guidance will be filled with many obstacles. The first and greatest obstacle one will face is the cursed Satan, who got our parents out of paradise after he tempted them to eat from the tree. He said, "Your Lord only forbade you this tree, lest ye should become angels or such beings as live for ever. And he swore to them both, that he was their sincere adviser" [Koranic verse; al-A'raf 7:20]. The evil devil sits in the path of goodness that the son of Adam may take to prevent him from obeying God, and to order him to disobey God Almighty. He sits in the path of Islam because he knows that by entering the religion of God, the servant will realize success in earning eternal life, as the Almighty said, "Only he who is saved far from the Fire and admitted to the Garden will have attained the object (of Life): For the life of this world is but goods and chattels of deception" [Koranic verse; al-Imran 3: 185]. Satan wants the son of Adam to die as a non-believer and to die in his polytheism so that he will be with him in Hellfire, and to listen to that Satanic lecture in the bottom of hell, burning in its fire. "And Satan will say when the matter is decided: 'It was Allah Who gave you a promise of Truth: I too promised, but I failed in my promise to you. I had no authority over you except to call you but ye listened to me: then reproach not me, but reproach your own souls. I cannot listen to your cries, nor can ye listen to mine" [Koranic verse; Ibrahim 14:22]. Like Satan, after telling people to disbelieve, and when they disbelieve, he tells them that he is not responsible for their deeds, and that he fears God. This is the case of man, when he follows the steps of Satan. He calls on him to disobey God, and he tempts him to defy the commands of the Almighty. But if he [man] falls, and follows his whim and is led by Satan, then he [Satan] will declare his innocence of his deeds and of his followers. Hence, no one follows Satan in this world unless he worships him, even if he denies it, and although he might not bow down, worship and pray for Satan, in reality he is worshipping Satan. This is what the Almighty said, "did I not enjoin on you, O ye Children of Adam that ye should not worship Satan; for that he was to you an enemy avowed. And that ye should worship Me, (for that) this was the Straight Way?" [Koranic verse; Yasin 36:60]

Therefore, all these non-believers, who are spreading corruption on earth, and who have oppressed the people and humiliated them, and who have taken mankind from light to darkness, and have forced them into disbelief, polytheism and clear misguidance, are but the soldiers and the friends of Satan, even if they boast shamelessly, and give themselves many titles. Therefore, Bush, Pervez, Mubarak, bin Saud, and Bu Tafliqah are nothing but soldiers of Satan. By fighting these criminals, we are fighting none but the soldiers of Satan, as the Almighty had said, "those who believe fight in the cause of Allah, and those who reject Faith Fight in the cause of Evil: So fight ye against the friends of Satan: feeble indeed is the cunning of Satan" [Koranic verse; al-Nisa 4:76].

Therefore, if the cunning of their president, their leader, their master and the chief commander of their armed forces is feeble, how [feeble] will be the cunning of his soldiers? The prophet, prayers and peace be upon him, said, "Satan sits in the path (of every goodness) that the son of Adam (may try to take)." Therefore, oh son of Adam, remember, when you want to take a path of obedience to God Almighty, be it through prayer, fasting, migrating, jihad or enjoining virtue, or preventing vice, remember that Satan is sitting in your path, because Satan swore to do so, and said, "Then will I assault them from before them and behind them, from their right and their left: Nor wilt thou find, in most of them, gratitude (for thy mercies)" [Koranic verse; al-A'raf 7:17]. Remember that you are in constant struggle, war and defense with Satan. Satan will try to prevent you first from obeying the Almighty God, and from the greatest obedience of all that will prevent you from entering into the religion of the Almighty.

In this light and this guidance, millions of people were deprived from this abundance and comfort. Who stood in between that and this, who is the one who decorated for them the pleasures, who made polytheism, non-belief, and going astray better to them in their eyes? Who is the one who reversed the darkness where they saw it as light, superiority, and culture and high class and prevented them from the light of Islam and rendered it inanimate, dead, old-fashioned, traded, fanatical, extremist, and severe? Who? It is cursed Satan. So remember, oh worshiper of God, that you are following Satan when you take the path of disobeying God, the Great and Almighty. "Verily Satan is an enemy to you: so treat him as an enemy. He only invites his adherents, that they may become Companions of the Blazing Fire" [Koranic verse; Fatir 35:6].

Then we are in a continuous war; our war is not limited and restricted to the war that we are carrying out with the enemies of God, the Great and Almighty, from Christians, Jews, apostates, rejectionists, Buddhists, Communists, and others, No! A part of our war, the largest part, is the war that we are carrying out against the master of those nonbelievers and with their president and leader; he is cursed Satan.

It is your duty, oh worshiper of God, to remember that you are always and at all times and in every obedience [to God] in a defense, dealing and fighting with this bitter enemy who wants for the human to be with him, in what? In the fire of hell. Satan sits in the path of the son of Adam in his paths. He sits for him in the path of Islam, and as we mentioned, this is the first step to salvation and success, for the human to embrace the religion of God, the Great and Almighty, meaning that he was salvaged, if he died and was in this religion, is saved from the punishment of God, Exalted He be. Either he was saved from entering the fires of hell from the beginning or he will not rest eternally in the fires of hell because those who died and were in Islamic religion do not stay in the fire of hell forever. He asked him, "Do you convert to Islam and leave your religion behind and the religion of your fathers and ancestors? How do you leave this religion that your father was raised upon it as well as your grandfather and your ancestors and move to a new religion? What is this religion? It is the Islamic religion, from which he wants to repulse you. And the excuses of what the predecessors were on is the excuse of the nonbelievers every time. They even said Nay! they say: 'We found our fathers following a certain religion, and we do guide ourselves by their footsteps. Just in the same way, whenever We sent a Warner before thee to any people, the wealthy ones among them said: 'We found our fathers following a certain religion, and we will certainly follow in their footsteps"[Koranic verse; Az-Zukhruf 43:22-23]. [Words indistinct] they copy those who were before them, but to ask themselves and stand up in pairs or alone to reflect, then their minds are dead. He said he stood for him in the path of Islam, and today we hear the same thing; we hear these Satanic excuses that stand between the people and embracing the religion of God, the Great and Almighty. So many say who are called to be Muslims and those who talk by their tongues but they are talking by Satan's tongue and they express the thoughts of Satan. How do you move to this religion that is long gone and we are in the era of civilization, in the era of advancement, and in the era of democracies, in the era of freedoms and in era and era. After that, what stands between you and God's religion, the Great and Almighty? Did God's religion prevent you from the industries? Did God's religion prevent you from traveling anywhere on earth? Did God's religion prevent you from advancing and technology? No! This is only it is the whispers of Satan and the dictations of his helpers. Therefore, this is not an excuse that passed and Satan mentioned for the predecessors who were before who used to worship a rock or trees. No! Rather, we hear it today from those who wear neckties, we hear it today from those who sit in the Parliaments, and we hear it today from those who call themselves intellectuals, the enlightened, the civilized, and others from these Satanic names who are the one who stand in between the people and embracing the religion of God, the Great and Almighty.

Islam is the religion of advancement, the religion of civilization, the religion of justice, and the religion of equality of rights. God has forbidden oppression upon Himself and rendered it forbidden among his worshipers. "Allah commands justice, the doing of good, and liberality to kith and kin, and He forbids all shameful deeds, and injustice and rebellion: He instructs you, that ye may receive admonition"[Koranic verse; An-Nahl; 16:90]. Is there anything higher and more supreme than these Islamic values of which the people are deprived, and they lived in this contemporary ignorance that burned their morals, beliefs, and manners and rendered them like pasturing animals who are like animals and even more astray? Islam is the religion of light; Islam is the religion of profusion; Islam is the religion of good life. Yes!

Then we must remember these valuable meanings ordered by the religion of God, the Great and Almighty. We should not be deceived or pay attention and not to be confused by what those criminals say. Those are the devils and helpers of Satan and soldiers for Satan. They drag the people with them, God's religion. [sentence as heard] They are those who invite at the doors of hell and those who answer them will be thrown in it and they do not care. Embracing God's religion means transforming the human from death to life. "Can he who was dead, to whom We gave life, and a light whereby he can walk amongst men, be like him who is in the depths of darkness, from which he can never come out?" [Koranic verse; al-Ana'm 6:122] Moving from disbelief to belief means being transferred from the deep black darkness to the light, as the Exalted He be, says: "Whoever rejects evil and believes in Allah hath grasped the most trustworthy hand-hold, that never breaks. And Allah heareth and knoweth all things. Allah is the Protector of those who have faith: from the depths of darkness He will lead them forth into light. Of those who reject faith the patrons are the evil ones: from light they will lead them forth into the depths of darkness" [Koranic verse; al-Baqarah 2:256-257]. Then brothers, we must be proud of this religion and hold on to it for our glory, strength, ability, and comfort in life and the hereafter is set by how much we hold on to it, our clinging and our pride in our creed and belief. "But honour belongs to Allah and His Messenger, and to the Believers"[Koranic verse; al-Munafiqun 63:8]. As for those who tumbled in the pits of being led astray, those who were deceived by names and the over decorated and the pleasures of this life, God, the Exalted He be, described them as: "They are only like cattle;- nay, they are worse astray" [Koranic verse; al-Furqan 25:44], "for the worst of beasts in the sight of Allah are those who reject Him: They will not believe" [Koranic verse; al-Anfal 8:55], "for the worst of beasts in the sight of Allah are the deaf and the dumb,- those who understand not" [Koranic verse; al-Anfal 8:22]. Those are: "They have hearts wherewith they understand not, eyes wherewith they see not, and ears wherewith they hear not. They are like cattle" [Koranic verse; al-A'raf 7:179].

Yes! Satan sat for the sons of Adam on this path to stand in between him and this great luxury, between this prevailing light that God, the Exalted He be, has honored us by for the entire humanity. Satan said: Do become a Muslim and leave the religion of you fathers and your ancestors? He disobeyed him and became a Muslim; this person has passed the first obstacle, and he passed it by the grace of God and His religion, the Exalted He be. "It is true thou wilt not be able to guide every one, whom thou lovest; but Allah guides those whom He will" [Koranic verse; al-Qasas 28:56]. Therefore, what you are into from belief, Islam, obedience, and untainted right and pure creed, is by what the grace of God, the Great and Almighty, bestows upon you, not by your experience or your expertise or your management or your inheritance, but rather "that is the Grace of Allah, which He bestows on whom he pleases: and Allah is the Lord of Grace abounding" [Koranic verse; Hadid 57:21]. "Those whom Allah (in His plan) willeth to guide, He openeth their breast to Islam; those whom He willeth to leave straying, - He maketh their breast close and constricted, as if they had to climb up to the skies" [Koranic verse; al-An'am 6:125]. "No soul can believe, except by the will of Allah, and He will place doubt (or obscurity) on those who will not understand" [Koranic verse; Yunis 10:100]. For God, the Exalted He be, is the one who gave you the permission to be in this religion. So cling to this luxury and thank God increasingly, the Exalted He be, for it.

"Then [Satan] sat in front of him in his path to Hijrah" [Hadith], but God, the prophet, may the prayers and peace of God be upon him mentioned in this hadith that the three great issues that have been repeatedly appended in God's Book [Koran]: Islam, Hijrah, and jihad. Jihad, as God Almighty, said: "Those who believed and those who suffered exile and fought (and strove and struggled) in the path of Allah,- they have the hope of the Mercy of Allah. And Allah is Oft-forgiving, most Merciful" [Koranic verse: al-Baqarah; 2:218]. And the Almighty said: "Those who believed, and adopted exile, and fought for the Faith, with their property and their persons, in the cause of Allah" [Koranic verse: al-Anfal 8:72] seek God's mercy and He is forgiving, and most Merciful.

And He Almighty said: "Those who believed, and adopted exile, and fought for the Faith, with their property and their persons, in the cause of Allah, as well as those who gave them asylum and aid"[Koranic verse: al-Anfal; 8:72]. And the Almighty said: "Those who believed, and adopted exile, and fought for the Faith, with their property and their persons, in the cause of Allah" [Koranic verse: al-Anfal; 8:72] is the greatest level [of faith] intended for God. So these three matters are repeatedly connected and mentioned numerous times in God Almighty's book: Islam, Hijrah, and jihad. Hijrah is the moving from non-Muslim countries to Muslim countries. This hadith, as proven by the repeated Koranic verses, shows us that parallel coexistence between believers and non-believers in one country is impossible and that a struggle will take place between the true believers and the non-believers and the misled, because Islam in itself does not accept untruthfulness and untruthfulness does not accept the true religion. For that reason the struggle between them [the believers and the non-believers] happens. So, either that the believers overcome the non-believers and defeat them and convert them to God Almighty's religion or they would pay the jizya "they pay the jizya with willing submission, and feel themselves subdued" [Koranic verse: al-Tauba 9:29], or that the non-believers overcome the believers or they [the believers] immigrate and leave the non-Muslim countries.

This is the Hijrah. For that reason God Almighty said in His great book: "the Unbelievers said to their apostles: 'Be sure we shall drive you out of our land, or ye shall return to our religion'" [Koranic verse: Ibrahim 14:13]. There is no alternative in dealing with the non-believers. For those who are today searching for a ground of coexistence between the believers, the monotheists, the forgiven [of non-belief], and the believers in the pure and true faith, with the non-believers who say that God is Trinity and that: "Uzair a son of Allah"[Koranic verse: al-Tauba; 9:30] or worship people, stone, trees and rats, they want coexistence between them, those who clash with the universal laws of the religion of God Almighty before clashing with His laws of the Sharia. God Almighty said: "The leaders, the arrogant party among his people, said: 'O Shu'aib! We shall certainly drive thee out of our city, thee and those who believe with thee; or else ye thou and they shall have to return to our ways and religion'" [Koranic verse: al-A'raf; 7:88]. So the prophet, may the prayers and peace of God be upon him, said: "Then Satan sat in front of him in his path to Hijrah" [Hadith], so he [the prophet] asked: will you immigrate? For he who immigrates is like a horse in countries. How do you leave this land, sky, country, job, family, wealth where you have lived, were raised and grew up, and move to a foreign and far away country and become restrained and unable to move around like a horse that is tied with a rope? For the immigrant is like a horse in a country. If you immigrate and leave behind your land and your country then your documents will become revealed and your passport will not help you to move around. You will be wanted by the American intelligence or the Saudi intelligence or the Iraqi intelligence or the intelligence of this or that [country].

So why do you make it hard for yourself when you are now in a safe, comfortable, and secure place with your children at your side and you teach and instruct them? When your wife is at your side and you have a stable job and a monthly salary? So why waste all this kind of life and move to hardship to a foreign country, to poverty, to pursuit, and to fear? But we say to Satan: 'you have lied for God Almighty has told us in His great book the opposite of what you said, for God Almighty said: "He who forsakes his home in the cause of Allah, finds in the earth many a refuge, wide and spacious, should he die as a refugee from home for Allah and His Messenger, his reward becomes due and sure with Allah. And Allah is Oft-forgiving, Most Merciful" [Koranic verse: al-Nisa; 4:100]. Therefore, he finds on earth a refuge which means that he can move around on earth, move from one place to the other, so if the place does not suit him he would move to another place, for God Almighty said: "O you servants of Mine who have attained to faith! Behold, wide is Mine earth. Worship Me, then, Me alone!" [Koranic verse: al-Ankabut; 29:56] Do not bind yourself to a land or a house or a job or a sky, for the earth is God's earth and that of his worshippers, the worshippers are God Almighty's worshippers and the issue is God Almighty's issue. You will die whether you are in your land or in the land of immigration. There is no difference. So why bind yourself [to a place]? And why should you follow the call of Satan who tells you that an immigrant is like a horse in the country of immigration? The immigrant who is true in the faith of God, the Great and Almighty, is on his way to heaven once he puts his foot at the doorstep of his house. God Almighty said: "He who forsakes his home in the cause of Allah, finds in the earth many a refuge, wide and spacious: Should he die as a refugee from home for Allah and His Messenger, his reward becomes due and sure with Allah" [Koranic verse: An-Nisa; 4:100]. Nothing will matter to him who seeks God and abides by the Sunnah of the prophet, may the prayers and peace of God by upon him, whether he dies or is killed, for heaven will be his reward without doubt. God Almighty said: "Those who leave their homes in the cause of Allah, and are then slain or die, on them will Allah bestow verily a goodly Provision" [Koranic verse: al-Hajj; 22:58]. And God Almighty said: "And if ye are slain, or die, in the way of Allah, forgiveness and mercy from Allah are far better than all they could amass" [Koranic verse: al-Imran; 3:157]. This is better than what they amass, better than the world and what it contains, better than your high ranks, better than your power, and better than your standing. Why do you waste this great blessing and hold on to the transient world which you are about to leave to those who are going to come next? So worshipper of God, this is God's extensive land and this is a call from God the Almighty who is calling upon you through His Holy Book, and these are His Almighty's commands that are being repeated to you day and night. So invest for the sake of God, immigrate for the sake of God, forget about the world that you are holding on to and clinging to and which led you to lowness and humbleness until we became subordinate to other people. Yes, for the prophet, may the prayers and peace of God be upon him, said: "But he disobeyed him and migrated from Makkah" [Hadith]. So overcome the second obstacle. I ask God Almighty to grant us the understanding of what we heard and make us of the true immigrants and to seal our immigration, for He is the Forgiver and Generous Provider.

All Praise is due to God. We worship Him, We seek His help and His forgiveness and we turn to Him in repentance. And we seek refuge in God from the evil within ourselves and our ill deeds. Whoever God guides cannot be led astray and whomever He leads astray, for him there is no guide. I testify that there is no god but God, with no partner, and that Mohammed is His servant, His messenger, and the chosen from among His creation. "It is He Who hath sent His Messenger with guidance and the Religion of Truth, to proclaim it over all religion, even though the Pagans may detest it" [Koranic verse; al-Tawbah 9:33].

The prophet, may the prayers and peace of God be upon him, said:

"Then he sat in front of him in his path of jihad and said, 'Why should you do Jihad? It will only exhaust your wealth and body. You will be killed, your wife will marry someone else, and your wealth will be divided to others. But he disobeyed him and went for Jihad'" [Hadith].

As we have previously mentioned, these three issues are very much connected in God Almighty's holy book: Islam, Hijrah and jihad. The prophet, may the prayers and peace of God be upon him, mentioned the third issue and that is jihad for the sake of God. He made it clear to us that Satan is exerting all his effort to stand between the worshipper and the bugle that summons to the field of jihad. Jihad as we know is the most difficult of adoration and that when God Almighty commanded it in His book, He said that a human being hates it, so He said: "Fighting is prescribed for you, and ye dislike it" [Koranic verse: al-Baqarah; 2:216]. Yes it is hateful to you because in it is loss of money and souls and the separation from beloved ones entails hurt, sadness, fear, hardship, devotion, and other feelings. God will reward his worshippers who follow His path in a manner equivalent to their hardship and struggle. Satan pretends to be a wise adviser, whose intention is your welfare; yet he is the same Satan who expelled our parents from heaven, pretending to be their adviser. He says you conduct jihad despite its hardship on the soul and wealth. Where do you want to go? Are you reckless, careless, and absent minded? You leave tranquility, peacefulness, and leisure to go to the crematories of Iraq or Afghanistan; you leave to go to the poor strange people in Somalia. This is the saying of the wise of this nation who are trying to prevent the youth from supporting their brothers, who are depleted by the war of the criminal crusader and others.

Yes, Satan came disguised as an adviser. We all know that the Satan does not talk to us directly but through agents who talk on his behalf. His agent who speaks to us might be a non-believing man or a misguided scholar who will misguide you or he might be a man pretending to be a wise scholar.

We should follow what is ordered in God's book. Did we not hear several times about scholars advising the youth against going to Iraq and they listed several points that prevent them from going to jihad.

From where did they come up with all these obstacles they adhere to and want the youth of the nation to follow? Almighty God ordered us clearly in his book in a manner that is clear to the scholars and the laymen. He said: "Go forth to war, whether it be easy or difficult for you, and strive hard in God's cause with your possessions and your lives: this is for your own good -if you but knew it!" [Koranic verse, al-Tawbah 9:41]

Do you think that this verse needs more clarification and philosophy? It is clear and obvious!

The scholars agreed that if the enemy attacks a Muslim country, it is the duty of the people in that country to fight that enemy until it leaves their country. If they are unable, delinquent, or negligent, the duty falls upon the neighboring country and if they are unable or delinquent, it becomes the duty of the next country and so on until the duty becomes the duty of the whole nation.

Were our brothers in Iraq able to defeat and expel the enemy? Were our mujahideen brothers in Afghanistan able to defeat and expel the enemy?

For almost six years they have been fighting; their fighting has its ups and downs and how many martyrs fell? How many hostages fell in the hands of the non-believers? How many women are widowed? How many children became orphans in the orphanages?

All these six years they are fighting, struggling, striving, and we still listen to those who try to prevent the youth from supporting their brothers. What kind of creed is this? What kind of faith orders a scholar to prevent the youth from supporting their brother? What is the value of these limitations?

The duty of jihad falls upon the Iraqi, since he is oppressed, and the Saudi, Kuwaiti or Syrian are not exempted from this duty because they are on the other side as the distance between them is two or three meters! Is this God's religion that you read in his book? Is that God's concern that you learned in the Sunnah of his prophet, prayers and peace be upon him?

When a Jewish man exposed the genitals of a Muslim lady, the prophet, peace and prayers be upon him, fought bin Qi'qa for her sake. This is God's religion, this is the creed that we learned, these are the orders we find in God's book. The philosophy, the temporary ideas, the deception, and the straying from the right path have no existence among us.

To the youth of this nation, strive hard to support your brothers in Iraq and Afghanistan; strive hard to support your brothers in Somalia, Algeria and Chechnya, and do not listen to Satan's suspicions that he dictates to his followers. Yes this is the religion of God we learned; whether they like it or not. He ordered us to strive hard and invited us to do so. "Satan said: Why should you do jihad? It will exhaust your wealth and body. You'll be killed, your wife will marry someone else, and your wealth will be divided to others" [Hadith].

These are the things to which humans adhere: "If your fathers and your sons and your brothers and your spouses and your clan, and the worldly goods which you have acquired, and the commerce whereof you fear a decline, and the dwellings in which you take pleasure - [if all these] are dearer to you than God and His Apostle and the struggle in His cause, then wait until God makes manifest His will" [Koranic verse, al-Tawbah 9:24].

Those who are discouraged, shaking and waiting, we warn if you do not support them, as you supported them before when there was a jihad against the Soviet Union, by God you will face the same situation like them [they will attack you also], the reward is according to the action. The prophet, prayers and peace be upon him, said: "But he disobeyed him and went for jihad" [partial Hadith].

This is the status of the Muslim; this is the status of those who have a clear and correct faith, those who are motivated because of their allegiance toward their brothers and God's sanctity. The prophet, prayers and peace be upon him, said: "It is a duty upon Allah that He shall enter the one who wins among them and dies, and if he gets killed it is the duty upon Allah that He shall enter him into paradise, and if he kills, it is a duty upon Allah that He shall enter him into paradise, and if he drowns, it is a duty upon Allah that He shall enter him into paradise, and if his animal broke his neck, it is a duty upon Allah that He shall enter him into paradise" [Hadith].

The following of Islam, emigration, and jihad is the guaranteed path to heaven. For those who are seeking heavens and God Almighty's pleasure; to those who are seeking to be among the companions of Mohammed, prayers and peace be upon him and his companions, the path to God and heaven is clear.

"O men! Verily, God's promise of resurrection is true indeed: let not, then, the life of this world delude you, and let not your own deceptive thoughts about God delude you" [Koranic verse, Fatir 35:5].

By God, a saving path which will lead to heaven or a destroying path leads to punishment in the hell fire. God worshippers, why are you wasting these blessings? Why are you wasting God's great gift? For the sake of a job, trade, money, or an ignorant person!

Expel those criminals who have spoiled the land and the worshippers, destroyed morals, and ruined characters. They built upon the ruins of the religion of God Almighty a credo of absconding, polytheism, loss, adultery, alcoholism and prostitution so that the children of Muslims have come to deny their religion, for there is no power or strength except through God. Oh God, glorify your religion, your book, and your believing worshippers. Oh God, glorify your religion, your book, and your believing worshippers. Oh God, grant your worshipping mujahideen timely victory and open their ways with clarity. Oh God, grant them support through your soldier, you, who possess soldiers in heaven and on earth. Oh God, they are poor, grant them wealth. Oh God they are hungry, give them sustenance. Oh God they are disbanded, grant them sanctuary. Oh God they are homeless, gather them. Oh God, defend them oh you who defends the believers. Oh God release your worshippers who are imprisoned and anguished. Oh God grant them ease from every tribulation and from ever worry a release. Bless them with that upon which they do not count. Oh God speed their delivery and their rescue. Oh God release them from the prisons of the Jews and the Christians and from the prisons of the apostates, the atheists, the Hindus, the Rejectionists and the Buddhists, for you are a Hearer, near and responsive. Oh God, prayers be upon the pinnacle of your creatures, Mohammed, and upon all his family and friends, and the last prayer is to thank God Almighty.

An Eid al-Adha Speech

Release Date: 22 Jan. 2008
Production Date: Unknown
Type: Video statement
Version: Translation

Praise be to God, we praise Him, ask for His assistance and His forgiveness, and we seek refuge to God from the evil of our desires and from our sins. Of those who are guided by God none can be led astray, and to those such as God rejects from His guidance, there can be no guidance. I witness that there is no god but God, the only, with no partners, and I witness that Mohammed is His worshiper and messenger and His chosen among His creation and His beloved. God sent him with the guidance and the religion of righteousness to prevail over all other religions, even if the non-believers disliked that. Prayers and peace of God be upon him and upon his entire family and companions and upon those who are following his guidance and method until Judgment Day.

Thereafter,

Lord, to you we offer great praise and gratitude that will satisfy you and please you. O God all the praise is for You, and all the thanks are for You, and all matters are returned to You. Praise to You, O God for guiding us. O God, praise to You, for making us the followers of the master of all people, O God praise to You, for the right religion He set as a law for us and the right path.

O Brothers of faith, this day is the day of al-Adha, it is the day of slaughter, it is the day of the great pilgrimage. This day is the day of one of Islam's great rituals in which the prophet, prayers and peace be upon him, introduced the methods, and set through it the laws that remained after him for his nation to follow and to adhere to. This day is the day on which our God saved Ishma'il, peace be upon him, from being slaughtered. This day is the day when loyalty to God prevailed over paternal emotions. So, praise be to God who made us the followers of Mohammed, prayers and peace be upon him, we commemorate his methods and hold on to his laws and we move along his guidance and follow in his footsteps.

O Brothers in Islam, on this day, I want to pause with a great Hadith from our prophet, prayers and peace be upon him, a Hadith that shows us the condition of the Islamic nation and how it should be at all times, from the day jihad was legislated until God inherits the earth and whomever is upon it. Narrated by Salamah Ibn-Nufayl al-Kindi, may God be pleased by him, he said: "I was sitting with the Messenger of Allah, prayers and peace upon him, when a man said: 'O Messenger of Allah! The people have held on to the horses and laid down their weapons and said: "there is no jihad for the war has laid down its burdens." So the Messenger of Allah, prayers and peace upon him, turned his face away and said: "They have lied. Now the time for fighting has come. There will never cease to be a group from my Ummah that fights upon the truth. Allah will deviate the hearts of some by way of them, and provide for them from them until the establishment of the hour and until the promise of Allah comes. Good shall remain in the manes of horses until the Day of Judgment'" [Hadith].

Yes, we talked a lot about jihad, they will say. Yes, we never stood at any time without talking about jihad, fighting, preparation and immigration. Such will be their words. Yes, the jihad is life for the Islamic nation, jihad is a continuing law that God, the Exalted He be, has legislated it and ordered for it in His book [Koran] and promoted it and praised the ones who performed it and cautioned from abandoning it. Yes, by jihad the religion will be straight, the laws will be preserved and the religion of the Almighty will be empowered. Our prophet, prayers and peace be upon him, when this matter was presented to him, "people held on to the horses"; meaning, they have humiliated the horses, in an account they let them go and abandoned them, they removed the weapons from it. When did this happen, O brothers? When did this matter happen? This happened when the Islamic state was at the glory of its establishment. This happened when the Islamic armies were striking east and west. And in the account that this Hadith was said after a conquest, which God had facilitated for His prophet, prayers and peace be upon him, the companions thought that after this conquest, this empowerment and this victory, there would be no need for jihad. The companion said: "O Messenger of Allah! The people have held on to the horses". And today there are so many horses that have been humiliated by the people and they have let them go. They abandoned and humiliated them. In fact, they renounced the entire Sharia of jihad, and not only the horses. "People have held on to the horses."

People have untied their horses and put down their weapons; they left their weapons; they stored their weapons; they filled the storerooms with weapons; and said no to jihad. Why? Because God has made us prosperous and has given us victory over our enemies. We have gained booty from them and we have broken their backs.

So what about today, then? If these words were uttered back in that day and age, when the Islamic state was at the heyday of its empowerment, when it was at the peak of its strength, when its leader and commander was our prophet, prayers and peace be upon him. This companion says: "People have untied their horses and put down their weapons, and said no to jihad."

How often do we hear this phrase: "No jihad"? There are those who say it openly, and there are those who may say it in passing, and there are those who may say it through their actions, and there are those who may say it by spreading doubts and discouragement and other means of which we hear and see. There are many ways to do so. It has been said by the friend, and it has been said by the enemy, and it has also been said by the neutral person. [They say:] No jihad. Why do you exhaust yourselves? Why do you tire yourselves? Why do you ensnare the Islamic nation into tunnels, of which there are no ends? How many youths have you ensnared in the holocaust of Iraq? How many youths have been entangled in the holocaust of Afghanistan? How many mothers have lost their sons? How many wives have lost their husbands? So then, what business is it of yours to conduct jihad? What business is it of yours to conduct jihad? What business is it of yours to get involved in this toil, in this stubbornness, in this strain? Why do you want to spend money? Why do you want to spill blood? Why do you want to face an enemy for whom you are no match? Do you know who [this enemy is]? It is America, who humiliated kings. It is America, who the nations of the East and the West have bowed down to.

No jihad. Do we not hear this phrase time and time again every day? And from whom? We hear it from scholars who should have been the heirs of our prophet, prayers and peace be upon him. They should have said to every person who spreads fear or discouragement: "They have lied. Now, now has come the time for battle." But unfortunately, at the head of the list of all the discouragers, and at the head of the list of the discouragers are those who are affiliated with knowledge. There is no power or strength except through God.

"People have untied their horses and put down their weapons, and said no to jihad." The war has relinquished its heavy load, meaning, the battles between us and our enemies have ended. So then, what need do we have for jihad? How can it be for us today, when our enemy is in our own house, when he has assaulted our pride, when he has stolen our money, defiled our sanctuaries, killed our children, imprisoned our elderly and our women, and in spite of all this we say: "No jihad, the war has ended."

Which war is this that has ended? Which war is this that the enemy has abandoned, while they are in the hearts of our lands? There you have Palestine, how many years has it been that the enemies of God; those who are the most humiliated and the most inferior of all of God Almighty's creations; how long have they been defiling her? There you have Iraq, how many years has it been since the enemies of God Almighty have been stealing her bounty, defiled all that is within her and assaulted the pride of the free within her? There you have Afghanistan, and what a situation there is in Afghanistan: enemy after enemy and conquest after conquest.

But in spite of all this we say: "No jihad, the war has ended." Then when will the war break out between us and our enemies? When will the war break out between us and our enemies? If we are not protective of our pride, then of what are we to be protective? If we do not leap for our faith, why then do we affiliate ourselves with the religion of God Almighty? Is it not the religion of God Almighty that tells us: "O ye who believe! What is the matter with you, that, when ye are asked to go forth in the cause of Allah, ye cling heavily to the earth?" [Koranic verse; al-Tawbah, 9:38] You are weighed down with this earth, you cling to her flowers and you bite down with your teeth and your molars on her riches.

Yes, and this is why all those have perished. The worshippers of the world will perish, as they were cursed by the prophet of God, prayers and peace be upon him. Abu-Hurayrah, in whom God is pleased, narrated that the prophet, prayers and peace be upon him, said: "Perish the slave of the dinar, dirham, qatifah [thick soft cloth], and khamisah [a garment], for if he is given, he is pleased; otherwise he is dissatisfied. Let such a person perish and relapse, and if he is pierced with a thorn, let him not find anyone to take it out for him". [Hadith]

Yes, and who is able to face all this? Who is able to face the worshipper of the world? Who is able to face the worshipper of the dinar, the dollar, glory, supreme power and the command of the unjust? He [the worshipper] will be faced by him about whom the prophet of God, prayers and peace be upon him, said: "Paradise is for him who holds the reins of his horse to strive in Allah's cause, with his hair unkempt and feet covered with dust: if he is appointed in the vanguard, he is perfectly satisfied with his post of guarding, and if he is appointed in the rearward he accepts his post with satisfaction". [Hadith] Yes, this person is not concerned with appearances, and is not concerned with his garments, and is not a person of high rank or a famous person. He is an unknown pious person who is not known by the media, and is not received by the satellite channels. As the prophet, prayers and peace upon him, had said: "If he asks for permission he is not permitted, and if he intercedes, his intercession is not accepted," [Hadith] as narrated by al-Bukhari.

This is the state of him who worships the world, and this is the state of him who has divorced [the world] and repulsed [the world] for the sake of God Almighty. So, paradise is yours, oh youth of Islam.

Rejoice for all the blessing and rewards bestowed by God Almighty. You are the poor emigrants [muhajirin]. The prophet, prayers and peace be upon him, said: "Some of you will die still having a need in their soul that is not fulfilled" [Hadith], as the prophet, prayers and peace be upon him, had said. There are many strong, fine and pious young men, who stand day and night to defend God's Muslim honor and sanctities, and yet no one hears or knows about them. They pass away with a need in their soul that is not fulfilled.

Abdallah Ibn-Umar Ibn al-Asy, may God be pleased with him, said that the prophet, prayers and peace be upon him, said: "The first group among my followers to enter paradise is the poor emigrants" [Hadith]. They are those who protect us against harm and when commanded they obey. They are those, who when commanded, they obey. This is their attitude and characteristic. Yes, they are the ones who obey when they are commanded.

"The Almighty, on Judgment Day, calls upon heaven and it comes with its ornaments and vanity. And, He says to it: Where are my worshippers who fought, strived and killed for my sake? Enter heaven without questioning. Then the angels come and prostrate in front of God, and say: Oh God we worship you night and day; who are they that you have preferred over us? The Almighty God says: They are my worshipers who fought and suffered for my sake. Then the angels come upon them and say: Peace be upon you, because you have persevered! "How excellent, then, this fulfilment in the hereafter!" [Hadith]. I ask God to let us be among them.

O Muslim youth in the east and west, who belong to the creed that is based on the proclamation: "God Himself testifies that He is the only Lord. The angels and the men of knowledge and justice testify that God is the only Lord, the Majestic, and All-wise" [Koranic verse; al-Imran, 3:18].

O Muslim youth in the east and west, who recite the book of God Almighty. "Who have attained to faith! Shall I point out to you a bargain that will save you from grievous suffering in this world and in the life to come?" [Koranic verse; As-Saff, 61:10].

O Muslim youth in the east and west, who listen to God calling you: "Go forth to war, whether it be easy or difficult for you, and strive hard in God's cause with your possessions and your lives" [Koranic verse; al-Tawbah, 9:41]. Is it not time to stand up and defend your religion? O Muslim youth, Islam is your religion and the creed of Islam is your creed. God is commanding you and inviting you to this great bargain [jihad] that is beneficial for you.

"God has bought of the believers their lives and their possessions, promising them paradise in return, and so they fight in God's cause, and slay, and are slain". [Koranic verse; al-Tawbah, 9:111]

Tighten your resolution against these non-believers, who filled the world with corruption. Kill them, and do not leave any of them. Do not listen to those who spread fear in your hearts. Do not listen to a discourager. Do not pay attention to a coward, who is engaged with life and has forgotten about the Hereafter. The prophet, prayers and peace be upon him, said: "They have lied. Now the time for fighting has come" [Hadith]. This is what the prophet of God, prayers and peace be upon him, had said. When did he say it? He said it, as we have said earlier, when the Islamic state was established, the Sharia was the law and the Islamic creed was pure, and God's enemies were scared and defeated. As for today, the Muslims are weak, the mujahideen are scattered and the children of Islam are suffering from humiliation, and have filled the prisons. The whips have torn their backs and their women have become slaves to the most despicable among God's creation.

In the prisons of Palestine you will find chaste women, who stood up to defend God's religion when the men were scared. Yes, this is our situation today, and we say: No jihad! What are we waiting for? Are we waiting for a miracle from heaven that will turn the world upside down so we can then establish an Islamic state, where Sharia is established, where our religion's foundations are erected, where our creed is widely spread? And then we will say: Now it is time for jihad!

Yes. All scholars have agreed that when the enemy invades a Muslim country, jihad becomes a duty of everyone in that country; if they are incapable or unable, then it becomes the duty of those surrounding [neighboring countries] them and able to support them. Where are we in this ruling? Where are we in this ruling, which we have opposed with our minds and with fabricated boundaries, which there is no truth in, or no value in, in the eyes of the religion of the Almighty, exalted He be. We have opposed it with our wills, with our philosophies, until we made the youth sit, and patted them on their shoulders and told them: "Sit down, rest, be smart," for that is not how things can be achieved. The enemy has overtaken fields and cities and we are still waiting for the victory to come from God Almighty. Victory is coming, God willing. I say these words, and I ask God for forgiveness for myself and for you, ask Him for forgiveness.

[al-Libi breaks for a moment]

Praise be to God, we thank Him and ask Him for support and ask for His forgiveness. We seek refuge with Him from the evils of our souls and from our bad deeds. He whom God has placed on the righteous path, no man will pull astray. And he who goes astray will not find the righteous path. I believe that there is no god but God, to whom there is no equal. I believe that Mohammed is His worshipper and His Prophet. Prayers of God be upon him, upon all his family and companions, and upon he who has followed his path and believed in his teachings until the Day of Judgment.

Thereafter: The prophet, prayers and peace be upon him, had informed us that the battle is upon us. And he had informed us of news that will remain true until the last of this nation battles al-Dajjal.

[Video jumps to new section]

It is the continuity of a nation that has been built upon truth, which is not harmed by those who oppose it or those who let it down. The prophet, prayers and peace be upon him, said: "Now is the time for battle and one group from my nation will always battle in truth." Not one individual, and not two individuals, rather an entire nation. Their battle is not in anger, it is not for patriotism, it is not nationalistic, it is not for mutual interests, and it is not for fabricated borders, but their battle is for the truth, and the most truthful truth is the [belief in the] monotheism of God Almighty. Jihad will not be jihad unless it is to raise the word of God Almighty. It is not to grant victory to partisanship, or to blood or the land, or anything else. But it is victory that will make the word of God prevail.

"One group from my nation will always battle in truth" [Hadith]. Therefore, O you worshippers of God, O you brothers of jihad in Palestine, in Iraq, in Afghanistan, conduct your jihad for God Almighty.

Realize victory for the zeal of the religion of the Almighty. Cast away pre-Islamic prejudice. Cast away pre-Islamic banners. For neither the land nor the blood [kinship], nor nationalism nor patriotism, moves us or motivates us. What really motivates us is "And fight them on until there is no more tumult or oppression, and there prevail justice and faith in Allah" [Koranic verse; al-Baqarah, 2:193]. O brothers in Palestine, cast away pre-Islamic prejudice, disown from the disbelief and its people and declare it openly and clearly: "We want to establish the religion of God Almighty; we want to implement the Sharia of the Almighty; we want to glorify the religion of God Almighty."

What is the use of evicting a Jew when after him an apostate takes his place, a secular Palestinian who is more vicious, filthier, and viler than this Jew we are fighting to cast out. The land does not glorify anyone; what glorifies a human being are his deeds. So if our deeds are for God Almighty, and our zeal is to the religion of God, the Exalted, then and only then, victory from God will descend. "If ye will aid (the cause of) Allah, He will aid you, and plant your feet firmly" [Koranic verse; Mohammed, 47:7]. Therefore, victory from God does not come by disobeying Him or by challenging Him, and not by violating His commands. Abu Musa al-Ash'ari narrated: "A man came to the messenger of God and said a man fights for war booty; another fights for fame and a third fights for showing off; which of them fights in Allah's cause? The Prophet said: He who fights that Allah's Word (i.e. Islam) should be superior, fights in Allah's cause" [Hadith].

Man has one soul, and when it comes out, it goes either to an everlasting eternity, or to punishment and hell [fire], so make your soul your dearest possession, make it a price for satisfying Almighty God, but not to satisfy some leadership, or a party or a country or an organization or a person or a friend or a relative or a stranger. It is a soul that you dedicate only for the sake of God and only God. "One group from my nation will always battle in truth" [Hadith].

Our prophet, God's peace and prayers be upon him, told us that the banner of jihad will not stop waving, its caravans will not stop, and its marches will not slow. Rather, it is continuous, since God sent His prophet, God's peace and prayers be upon him, and legislated to him this religion, and until Judgment Day, as mentioned in this Hadith. It is a good tiding from the prophet, God's peace and prayers be upon him, which we should take and be content with, knowing that God is with us, and that a good outcome is for the pious, and that enforcing this religion is either done by us or through us by being the fuel that moves the jihad wheel. The empowerment of the religion is coming, no doubt. Ubay Ibn-Ka'b narrated, and said, the prophet of God, peace and prayers be upon him, said: "To the Ummah, bring good tidings of facilitation, sublimity and glory for the religion, and empowerment and victory in the lands. Whoever carries out the deeds of the Hereafter in this world does not have a reward in the Hereafter" [Hadith]. Our deeds must be purely for the Almighty, exalted He be. We must not direct them at anything but to Almighty God. "There will never cease to be a group from my Ummah that fights upon the truth. Allah will deviate the hearts of some by way of them, and provide for them from them until the establishment of the hour and until the promise of Allah comes. Good shall remain in the manes of horses until the Day of Judgment" [Hadith].

We ask the Almighty God to glorify His religion, enable His Sharia and support his believers. O Lord, help your mujahideen believers to realize utter victory. O Lord, grant them clear success. God grant them victory in Iraq; O God, support them in Afghanistan and Pakistan. O God, support them in Palestine. God destroy Your enemies, who are fighting Your religion and torturing Your followers. God, destroy the Arab and non-Arab tyrants, kill them in numbers, finish them off, and do not spare anyone. God show us what will please our eyes and satisfy our anger towards them. God destroy them. God show us their black day. God defeat the demoralizing hypocrites. God punish the criminal hypocrites. God claim the demoralizing hypocrites. God, harm them for the sake of your believers. To You we come with our weakness, O God, and to You we bring out complaints, so help us. To You we bring our needs. God, you are our Lord and there is no god but You. You are our God, and there is no god but you. O Lord, do not make us depend on anyone but You, even for a moment. God, do not let us depend on ourselves, or anyone else but you, even for the twinkling of an eye. God make this day a day of victory, conquest and empowerment for Your believers. God assist the needy; God assist the needy; God assist the needy.

God release our captured brothers in the prisons of the Jews and the Christians, and the prisons of the apostates and the atheists, and in the prisons of the rejectionists and the Buddhists. God help them in their afflictions, and help them in their hardships. God, give them their daily bread and return them to their families safe, secure and victorious. God support Your religion and glorify the religion and the book by them, O Lord of all creation. God pray on the best of Your creation, Mohammed, and all his family and companions. Our last supplications are to praise God, the Lord of all creation."

The Companion

Release Date: 2 Mar. 2008
Production Date: Unknown
Type: Audio statement on video
Version: Translation

In the name of God, the merciful, the compassionate. All praise is due to God, and prayers and peace be upon God's messenger and on his family, companions, and allies. O nation of Islam, men of sacrifice, martyrdom, and courage, may God's peace and blessing be upon you. This is how the school of Jihad taught us and this is how we drank and drink from its incessant spring. These are continuous lessons, hardship and ease, joy and sorrow, victory and defeat, lessons and tears, receiving men and bidding farewell to heroes. Success lies in the hope for a reward after patience and firmness on the path of God based on faithfulness to Almighty God. "And slacken not in following up the enemy: If ye are suffering hardships, they are suffering similar hardships; but ye have Hope from Allah, while they have none. And Allah is full of knowledge and wisdom." [Koranic verse, al-Nisa, 4:104]

Days of varying fortunes between us and our enemies are by turn. We suffer and they suffer, we are pained and they are pained, we endure and they endure, but the end in this life and in the hereafter is best for the righteous ones. "If a wound hath touched you, be sure a similar wound hath touched the others. Such days (of varying fortunes) We give to men and men by turns: that Allah may know those that believe, and that He may take to Himself from your ranks Martyr-witnesses (to Truth). And Allah loveth not those that do wrong." [Koranic verse, al-Imran, 3:140]

Among the greatest things we have tasted and seen in the arenas of jihad, of the bitter pains and heavy sadness is the parting of the beloved ones, the farewell to companions, and the absence of friends. Our sadness is not only for their killing and departure, for we hope that they will have the honor they have sought and about which they were concerned. However, we are sad because we were left behind them. I swear by God, I wished I had left with my companions when they were bade farewell and buried in their pure tombs.

There are three things when patience becomes difficult and which startle any wise person, these are a forcible departure from a country you love, separation of brothers, and the loss of a beloved one. [Verses of poetry]

We have never known the true friendship, the meaning of the brotherhood in faith, and the consolidation of pure amity except in the arenas of jihad, where souls meet while flying in the sky of longing to God, and where blood is mixed while flowing generously over the land of loftiness and competition over good deeds manifests itself in the best of its images and loftiest forms, especially since there is no fierce struggle over worldly things or perishing pleasures. It is nothing but the spirit of faith, quest for paradise, and concern about martyrdom. Only a deprived person can miss the pleasures of life in this world, and may God protect us from that.

O nation of Islam, today, we announce to you the death of one of the courageous heroes of Islam and one of its unique and great leaders who were worried about the situation of their nation and what it is currently experiencing in terms of weakness after strength, humiliation after pride, and lowliness after loftiness. He struggled with life in order to rescue the nation from its decline, save it from its predicament, and bring it back to its past glory. Thus, his loss was a real loss and his absence was a real absence.

Upon your life, heavy loss is not the loss of money or the death of a goat or a camel, but heavy loss is the loss of a free man after whose death many people die. [Verses of poetry]

O beloved nation, we announce to you the death of one of the men who have recorded their stands in the annals of history with the endurance of men, the courage of lions, the firmness of leaders, and the humility of the faithful to another faithful. His end was the best. This is what we believe and God knows better. “And say not of those who are slain in the way of Allah. They are dead. Nay, they are living, though ye perceive (it) not.” [Koranic verse, al-Baqara, 2:154] We announce to you the death of a living example of those made by jihad and shaped by faith and who are known of their pride in their religion. Thus, they rejected humiliation, renounced lowliness, and discarded submission and adhered strongly to the Koran and thus, God elevated them by it. “If any do seek for glory and power, to Allah belong all glory and power.” [Koranic verse, Fatir, 35:10]

We announce to you the death of one of the lions of wars, protectors of truths, companions of swords, and causers of the death of enemies, mujahid Sheikh, educator, leader, and courageous lion, Abu-al-Layth al-Qasimi, may God have mercy on him, accept him among martyrs, elevate his status in the loftiest paradise, and make him the companion of prophets, saints, martyrs, and righteous people, who are the best of all companions.

On our great loss, we say that the heart is grieved and the eye is full of tears, but we only say what pleases our Lord. O Abu-al-Layth, we are grieved because of your departure. "Be sure we shall test you with something of fear and hunger, some loss in goods or lives or the fruits (of your toil), but give glad tidings to those who patiently persevere, who say, when afflicted with calamity 'To Allah We belong, and to Him is our return.'" [Koranic verse, al-Baqara, 2:155-156] To Allah we belong, and to him is our return.

I complain to God, not to people, about his loss and about the pain of sadness in the heart. [Verse of poetry]

Following a march full of sacrifice, offering, persistent action, and continuous jihad, which are all surrounded with great loftiness, shrewdness, mental acuteness, long and various experiences, and noble manners, our courageous sheikh and lion bade farewell to the arenas of jihad, preparation, and steadfastness. In fact, he bade farewell to life as a whole to hand over the flag to those after him while he seemed to say: "For the like of this let all strive, who wish to strive." [Koranic verse, al-Saffat, 37:61] We believe that he was among those who have been true to their covenant. "Among the Believers are men who have been true to their covenant with Allah. of them some have completed their vow (to the extreme), and some (still) wait: but they have never changed (their determination) in the least." [Koranic verse, al-Ahzab, 33:23]

He had never deviated from the path of the guided ones, but he had always been true to his covenant with God and he had never changed,

Death came to him with a fixed glance, but he did not turn his back and said to it: It is you what I wish and so be prepared. [Verses of poetry]

Yes, he bade farewell to those arenas which he was familiar with and which were familiar with him. He loved them and they loved him. With his pure blood he wrote what he had previously wrote with his friendly ink when he said about his brothers who were martyred before him: O companions, we have received your message and you have presented you argument and paved the way for those after you with the purest thing you have, and we can only say: "It is enough for martyrdom to be the bond of kinship among us if we were not united by the bonds of kinship, and no matter how happy we were with you, we still have a pledge to fulfill." [poetry] By God, we are fed up with life after you, its sweetness became bitter and its delicious things now cause choking. It is only one of the rounds of truth against falsehood until God grants victory to religion and allows us to join you in the loftiest paradise, God willing.

This is how words come out of the hearts, which are true to their love, which have strong faith, which are certain about the soundness of their path, and which are faithful to their lovers and companions. They come out with the spontaneity of lovers and the smoothness of the longing ones. They are void of affectation, ornamentation, ambiguity or rhymed prose. They infiltrate into hearts to settle deep inside them and activate their hidden things, and they cannot be prevented by anything.

Our sheikh spent 19 years in the various arenas of jihad. He attacked, assaulted, and fought. He was patient and stood fast. He waged battles, was used to all kinds of conditions, embarked on risky undertakings and disdained their threats and stormed hardships and cared less about them. He surmounted all difficulties by his will and determination and made them ease by his loftiness and wisdom. I can hear the echo of his voice reverberating between the mountains of Afghanistan, the deserts of Libya, and the prisons of al-Saud.

I would rather make what is difficult easy or die, for hopes can only be achieved by the patient ones. [Verse of poetry]

This in order to explain to his nation the path of glory, the course of enabling, and the road to attain noble things, especially since imamate in religion can only be attained through faith and patience. "And We appointed, from among them, leaders, giving guidance under Our command, so long as they persevered with patience and continued to have faith in Our Signs." [Koranic verse, al-Sajdah, 32:24]

It is impossible for anyone to attain noble things without hard work. [Verse of poetry]

He couldn't bear living under the humiliation of tyrants and the rules of pre-Islamic paganism while hearing God's saying: "He who forsakes his home in the cause of Allah, finds in the earth Many a refuge, wide and spacious: Should he die as a refugee from home for Allah and His Messenger, His reward becomes due and sure with Allah. And Allah is Oft-forgiving, Most Merciful." [Koranic verse, al-Nisa, 4:100] He heeded the call for jihad and flew to the arenas of glory and sacrifice soul before body. He tells anyone who clings to life and its vanities and who finds it difficult to leave his family and children: "Your departure for the sake of attaining noble things is not considered a travel, but staying behind with humiliation is travel." [poetry]

Yes, he was determined, just like others who have earnest concern about religion and faith and longing for God, to knock the doors of paradise at all times. He did not pay attention to those who justify their staying behind by the allegations of wisdom, cover their concern about life in this world by patience and wisdom, while reading God's saying: "Unless ye go forth, He will punish you with a grievous penalty, and put others in your place; but Him ye would not harm in the least. For Allah hath power over all things." [Koranic verse, al-Tawbah, 9:39], and "Say: "The Death from which ye flee will truly overtake you: then will ye be sent back to the Knower of things secret and open: and He will tell you (the truth of) the things that ye did!" [Koranic verse, al-Jum'ah, 62:8] He seemed to say to everyone who fails to perform jihad: "Some people say that my opinion about courage is ignorance, but I believe that when they flee death they are more ignorant, one cannot flee death, and if I avoid death in the battlefield, then I will die somewhere else." [poetry]

God ordained for him the honor of pa rticipating in fighting the greatest empires of tyranny and injustice in this age, the Soviet Union, which has become something of the past, and the mother of all evils, America, whose civilization will turn into a waste land and its power will turn into mirage, God willing. He became strong from the lessons of jihad, in words and deeds. He gained experience from developments to the point where he managed to probe their depths and to realize their ins and outs while he seemed to say: "I mastered harms until I made them say: Has death died or fright was frightened?" [poetry]

The storm of the cross blew in its modern war and mobilized its forces and its agents and advanced to the fortress of Islam, from land, sea, and air and the weaklings began to spread the news and despair and to mock the small group of the people of faith. Very few people, including our sheikh, may God have mercy on him, stood, along with their true reliance on God, disdain of falsehood, and pride in their religion and willpower, following the example of their ancestors in good and bad days and saying: "Men said to them: 'A great army is gathering against you': And frightened them: But it (only) increased their Faith: They said: 'For us Allah sufficeth, and He is the best disposer of affairs.'" [Koranic verse, al-Imran, 3:173]

Every time someone suggests to them the fear of the enemy and its power, they had deep in their hearts, which are full of faith, God's saying: "It is only the Evil One that suggests to you the fear of his votaries: Be ye not afraid of them, but fear Me, if ye have Faith." [Koranic verse, al-Imran, 3:175] Thus, they served as their nation's shield to deter calamities, fighting the sword with a sword, repulsing injustice by decisiveness and shedding the enemy's blood in return for their blood until they healed their breasts and those of the believers. They taught the worshippers of the cross and their hinchmen how men act in the battlefield. They torn their bodies, shed their blood, orphaned their children, and made them swallow chokes of pains, which are still chasing them wherever they go. They have taught them lessons, which their generations will remember one generation after another. So, what is the harm of falling martyrs after this? "Say: Can you expect for us (any fate) other than one of two glorious things- (Martyrdom or victory)? But we can expect for you either that Allah will send his punishment from Himself, or by our hands. So wait (expectant); we too will wait with you." [Koranic verse, al-Tawbah, 9:52]

Our sheikh, may God have mercy on him, knew the road to paradise and adhered to it. He displayed patience toward its adversities because he knows that the road to it is surrounded with adversities. We believe and God knows better, that the saying of the prophet, may God's peace and blessing be upon him, that "every time he heard a cry he flew to it expecting death" applies to him. Every time he was in distress he made it easy by his quest for the best end. This distress was kept away from him by what the patient ones were promised, which included in God's saying: "Those who patiently persevere will truly receive a reward without measure!" [Koranic verse, al-Zumar, 39:10] So, O venerable sheikh, sleep happily. By God, we hope that you have won one of two glorious things [martyrdom or victory], or even the best of the two glorious things, martyrdom in the cause of God, which is the ultimate goal and the long-cherished wish, eternal happiness, pleasant life, joy and delight, comfort and easiness, and greater reward from God. We know that you had been very eager about it and serious in seeking it, moving among the things that lead to it, and intercepting it in the places where you can find it. Congratulations to you for you have won what you wished and gained what you have expected, God willing.

We tell you and your brothers who were reddened by their blood along with you, of our beloved ones and companions, that we will continue to pursue the march, keep the pledge, fight and renounce the enemies of religion until we win what you have won or God decides between us and the infidels. Our sadness over your departure will not weaken our adherence to your path. We will not cry like the weak ones, but it is blood for blood and destruction for destruction. "And soon will the unjust assailants know what vicissitudes their affairs will take!" [Koranic verse, al-Shu'ara, 26:227] "And the close of their cry will be: "Praise be to Allah, the Cherisher and Sustainer of the worlds!" [Koranic verse, Yunus, 10:10]"

I am Not a Deceiver, Nor Will I Allow Someone to Deceive Me

Release Date: 9 Mar. 2008
Production Date: Unknown
Type: Video statement
Version: Translation

In the name of God, the Compassionate, the Merciful. Praise be to God, and prayers and peace be upon the messenger of God, and upon his family and companions, and those that followed him.

Nation of Islam, God's peace, mercy, and blessings be upon you. The band of truth, the sect of jihad is still engaged in an ongoing and ever changing struggle with the enemies of Islam from various faiths and sects including hateful Jews, arrogant Christians, and treacherous and defeated apostates. With each day that goes by, the battle gets more fierce, not to mention the number of fronts and the range of tactics used in it. No sooner do the enemies of our religion discover that they lost a battle against Islam and Muslims in general and mujahideen in particular than they are led by their evil minds to look for a new door that would allow them to achieve their only objective which is to make us abandon our religion and go backward after we receive guidance from God. We were told by the one who knows what is in their hearts [in the Koran], when the Almighty said: "[Your enemies] will not cease to fight against you till they have turned you away from your faith, if they can" [Koranic verse; al-Baqarah 2:217]. And the Almighty also said: "Out of their selfish envy, many among the followers of earlier revelation would like to bring you back to denying the truth after you have attained to faith - even after the truth has become clear unto them" [Koranic verse, al-Baqarah 2:109]. God Almighty also said: "They would love to see you deny the truth even as they have denied it, so that you should be like them" [Koranic verse; al-Nisa 4:89].

There are only two clear paths – unambiguous and unmistakably distinguished – the path of truth and guidance chosen for us by God and which were guided by Him, and the path of inclinations, desires, and pleasures chosen and promoted by those deviants. God Almighty said: "God wants to make [all this] clear unto you, and to guide you onto the [righteous] ways of life of those who preceded you, and to turn unto you in His mercy: for God is all-knowing, wise. And God wants to turn unto you in His mercy, whereas those who follow [only] their own lusts want you to drift far away from the right path" [Koranic verse; al-Nisa 4:26-27]. All of their efforts are directed toward stripping us of our religion and making us move away from our Islamic laws so that we become their equal in non-belief. Every path they take and every way of life they promote or try to tempt us into is intended for this same objective, may God protect us from their treachery. We must therefore always remember this clear Koranic truth as we fight on every battlefront in our war with them. We must be very conscious of it. We cannot let ourselves be deceived with their duplicity or fall for their colorful language or be enticed by their claims and deception. We must be aware, awake, and alert on all fronts and wherever we are facing them, including in the military and on the intellectual, media, economic, or moral values fronts.

Regardless of how nice they pretend to be and no matter how often they claim to be concerned and how many slogans of tolerance and friendship they wave, those non-believers hide the truth [about their intentions and their attitudes] that is entrenched in their hearts and cannot be removed and which was mentioned in the words of God Almighty when He said: "O you who have attained to faith! Do not take for your bosom-friends people who are not of your kind. They spare no effort to corrupt you; they would love to see you in distress. Vehement hatred has already come into the open from out of their mouths, but what their hearts conceal is yet worse. We have indeed made the signs [thereof] clear unto you, if you would but use your reason" [Koranic verse; al-Imran 3:118]. They, may God destroy them, do not spare any effort to bring us corruption and to cause us hardship and difficulties. They tried everything and every means to achieve that. They used their hands, mouths, pens and everything that is within their reach, but still, the hatred, enmity and spite they hide within their hearts is much more than what we can ever see, hear or touch. Brushing off or ignoring this truth will have serious consequences and deadly results for the heedless and the simpleminded. The enemies of Islam will deceive them with their false smooth façade but will serve them the poison at the right time. They will then throw their cadavers in the dark abyss of non-belief from which they will find no return unless God wills it. That is the requital of the evildoers. God Almighty said: "O you who have attained to faith! If you pay heed to some of those to whom revelation was vouchsafed aforetime, they might cause you to renounce the truth after you have come to believe [in it]" [Koranic verse; al Imran 3:100]. God Almighty also said: "O you who have attained to faith, if you pay heed to those who are bent on denying the truth, they will cause you to turn back on your heels, and you will be the losers" [Koranic verse; al-Imran 3:149].

Nation of Islam, What caused me to speak about this matter is what was thrown at us by the Egyptian security forces, the master of deception in this age, and what they called "The Rationalization of Jihad in Egypt and the World Document," which was attributed to Sheikh Sayyid Imam. The one who reads this titled and its content bearing in mind that it was a product of those [security] services can only think what the poet said:

"In Egypt there are many laughable things,

Except that laughter feels like shedding tears." [poetry]

Since when have the instruments of persecution and killing, the masters of terror and torture, the houses where the Lord is insulted and the noble Koran is mocked and the laws of God are fought, since when have these become concerned about guiding the people to religion, let alone guiding the mujahideen to the right path to establish, safeguard, and ensure this absent religious duty is properly followed, that is the duty of jihad. But as was said: "If you have no decency, you are free to do as you wish." The mere fact that this document came out from the dungeons of deception and treachery shows you its true nature and makes you ponder the motives behind it and makes you stop and think about every word in it and every other line and every symbol it contained. Do not let the exploitation of the writer's name used to promote the document fool you. Do not bother about questions such as why he changed positions or why he retracted, for that is not the objective. The issue is much bigger and more dangerous. Take the muhkam and leave the mutashabih. What is muhkam about those despicable people is the fact that they have a deceptive nature, a determination to spread corruption, an earnest desire to destroy Islam, and the willingness to spend time and money to enslave people to their false gods willingly or by force. The legacy of their past, their present, their statements, and their actions all testify to their crimes and expose their deceptive nature. They are the ones who denied the truth and were immersed in iniquity. They are unjust, corrupt, and agents of corruption. The misguidance and falsehood in their document is too weak to stand up to the clarity of the truth or to resist the brightness of the straight path. They are like the Jews about whom God said: "And some of the followers of earlier revelation say (to one another): 'Declare your belief in what has been revealed unto those who believe (in Mohammed) at the beginning of the day, and deny the truth of what came later, so that they might go back (on their faith)'" [Koranic verse, al-Imran 3:72]. This is what those criminals are doing. It is what they want also. They pretend to be compassionate, lenient, and concerned offering advice and guidance in order to lure mujahideen and trap them in their snares and polytheism and in order to win the hearts and minds and to gain the trust, but once they achieve their objective, they will show their fangs, pronounce their non-belief, reveal what they were hiding, and pounce on jihad and the mujahideen and their supporters, catching them by surprise after besieging them. Be warned and alert.

We all know that if this document was adopted by the Egyptian security service and was signed by one of their officers and the heads of non-belief in it and was presented to the mujahideen as advice from the security service to ensure the success of jihad and to strengthen it and purify it, everybody would have laughed at them and they would have been subject to ridicule from the whole world and every Muslim would have regarded it as a trick. What makes it different then? What alters its nature? Is it because it came out under the name of somebody? The only way they could give it [the document] legitimacy is by circulating it in the this false cloth; that is, issuing it under the name of the Sheikh Sayyid Imam Abd-al-Qadir Ibn-Abd-al-Aziz, may God give him and all Muslim prisoners solace. The poet said it beautifully about those in similar situations [in captivity]:

"Wolves might play sheep sometimes,

But be aware for they might turn into wolves again.

The wolf is most deceptive when

It is dressed in little lambs clothing.
[end of poem]

Nation of Islam, my mujahideen brothers," [poetry]

The pharaoh-like descent on which these criminal, oppressive apparatuses were built, which is founded on the basis, "Their male children will we slay; (only) their females will we save alive; and we have over them (power) irresistible" [Koranic verse; al-A'raf 7:127], and on the basis "If thou dost put forward any god other than me, I will certainly put thee in prison!" [Koranic verse; Ash-Shu'ara 26:29] and on the policy "Does not the dominion of Egypt belong to me, (witness) these streams flowing underneath my (palace)? What! see ye not then?" [Koranic verse; Az-Zukhruf 43:51] is the one that guided them to this Satanic path, on which their cursed forefathers preceded them by His saying: "Pharaoh said: I but point out to you that which I see (myself); Nor do I guide you but to the Path of Right" [Koranic verse; Ghafir 40:29].

And because the devils were the ones who stirred the pharaoh and they stir the pharaohs of the era, then their titles match and their actions are alike and their crimes are the same too. So this is the tyrant pharaoh who had the audacity against God, the greatest, to dare and torture His servants in the most horrible torture and corrupted the earth with the utmost corruption and the one who became an example in arrogance and tyranny to act strong and with superiority, describing his invitation by the exposed deviation from what is right and the composed disbelief by his saying "Nor do I guide you but to the Path of Right" [Koranic verse; Ghafir 40:29]. The Egyptian security system that discharged out to us this document – the document of deviation from the right path, of betrayal, of disgrace, and of accepting the reality and surrendering to the oppression and the oppressors – call it the "Document of Rationalization of Jihad." For that damned [pharaoh] calls his invitation guidance and those criminals call it guidance also. The one before them said the same; their hearts are alike: "We have indeed made clear the Signs unto any people who hold firmly to Faith (in their hearts)" [Koranic verse; al-Baqarah 2:118], so glory to God! "Is this the legacy they have transmitted, one to another? Nay, they are themselves a people transgressing beyond bounds!" [Koranic verse; Az-Zariyat 51:53]

The tyrant pharaoh, who accustomed his people with the policies of superiority, arrogance, degradation, and enslavement, did not renounce his disdain and come down from his superiority to address his people and to fill them with his delicate speech, claiming that he was concerned about them and about their guidance and making efforts in guiding them until the argument of the prophet of God, Moses, peace be upon him, left him dumbstruck, and it shook the people's confidence in him [the pharaoh] so he feared that the matter may go out of his hands and he feared that the faith would infiltrate into their hearts from what they saw of the obvious signs and the marvelous miracles that were carried out by the prophet of God, Moses, peace be upon him. So he [pharaoh] sought the methods of trickery, cunning, and deception, and he wore the dress [pretended] of the one who guides to rationality, the sympathetic advisor, and the one who invites to the path of guidance.

The story of the document of guidance that was born [created] disfigured in the hatcheries of the Egyptian security systems brings back this same scenery in its aspects and contents so that we know that the righteous battle against the wrong is one and that the multiplicity of its manners and the difference of its tools and the advancement of its instruments do not mean, in any way, the change of its reality or that it has become extinct. Therefore, we say with all pride and confidence that one of the greatest gains of holding fast to the jihad path and to stay steadfast on it is to force these arrogant, tyrannical, superior systems to submit, while they are unwilling and compelled rather than addressing the people of jihad in this logic which is new to them, the logic of evidence, proof, and argument, even though they lied in their allegations and were led to deviate from the right after they did not deal with them [mujahideen] or talk to them, other than with one language; that is the language of skinning them, breaking their bones, violating their honor, burning them by the fire, burying them [left for long periods] in prisons and overpowering their women, their children, their families, and their relatives. So here they are today, first by the grace of God and then by what God opened to the mujahideen of being steadfast on the righteous method and the continued the consecutive strikes that fell on their masters and humbled their superiors and overpowered their lords. Here they are today, discovering that the language of the whips is not useful and that burying the living in the prisons does not stop the battle and that the military courts and make-believe trials do not add to the jihad and the mujahideen anything but steadfastness and pride. So they stopped, even though for a short while, at the pharaoh-like method: "Nor do I guide you but to the Path of Right" [Koranic verse; Ghafir 40:29].

So what rationalization or guidance does this document of betrayal invite? Is it the rationalization that says to the mujahideen and the Muslims to stop [jihad] and let us shed the blood and rip the bodies apart whenever we want, or is it the guidance that tells them to "be submissive and lenient and lower your wings and let us rob from your riches whatever we please" or is it the guidance that tells them to "mind your own matters and dedicate yourselves to your own troubles and leave us to cause mischief in the land" or is it the rationalization that says to them "hold on to your lives and run after gaining your food and devote yourselves to seek for your daily bread and let us lead you with our desires and burn you in the inferno of our decisions" or is it the rationalization that states "what do you have against the occupying Christians and the man-slayer Jews and the domineering apostates; so if they occupy your land, surrender, and if they overpower you, be submissive, and if they enslave your women, endure, and if they rob your wealth, then do not condemn it." Then what precaution is this, whose springs erupted in the hearts of the tyrants, the violent, and cruel hearted and whose people prepared the right climate for it and spent large amounts of money and dedicated numerous media agencies and carried out consecutive interviews to convey the words of advice and the phrases of guidance and the meanings of compassion toward the mujahideen? And who awakened those executioners after a series of their hideous deeds and the path of their crimes that is impossible to forget or erase from the memory of history, to show us, the mujahideen, what the Islamic law is and that we have to respect and take hold of it and move according to it in our jihad against our enemies? Are we this stupid, foolish and dumb to believe that the Egyptian Government and its security systems have changed overnight to become the al-Azhar University, which was once prosperous- for us to grab what it throws at us, resting assured that whoever gives us the fatwa from there or advises us, or guides us are impartial and sympathizing scholars, so we take what they say as if it were pure honey or agreeable milk? But rather the claims are from those who propagate the document of betrayal, whose contents are recorded, to the one to whom it was attributed by his total choice and free will far from pressure, force and restraint. Here it is our right as reasonable people to ask an innocent question which is: If that document expresses the belief of its writer to whom it is attributed, then why are security systems keeping him [Sayyid Imam] behind bars? And why are they saving him in their prisons, since he is able to extend to them services that exceed what they want from him? So let them send him to the fields of jihad and the lands of the outposts so that he can argue with those mujahideen verbally to show them their mistakes and to caution them from the evil of the pitfalls that they alleged.

So let the Egyptian security systems take the initiative for this step to complete its path of advice and rationalization to the jihad and the mujahideen and they will get what they want. "Ah! what an evil (choice) they decide on" [Koranic verse; An-Nahl 16:59]. As well, it is known that those to whom the document of rationalization was attributed – whether the attribution is true or not – are among those who know best the truth about these systems of the Egyptian pharaoh-like system, and they are most aware of their treatment with those whom they suspect of violations [think they might have the slightest affiliation] to the Egyptian pharaoh-like government whether they were from the Islamists or from others, aside from being from the mujahideen. Those were also leaders in the jihadist groups. Their books, their foundations, their fatwas, their research are all still proof of their stand toward jihad and all their literature shows their stance toward it. For those in such positions, everyone knows how the Egyptian intelligence system and others deal with him and they know that the types of torture and the varieties of severe punishment that the satans of the era have reached were all part of his share and were waiting for him. So what caused these spiteful systems to refrain from all of that and replace it with the policy of openness and ease and provide the means of rationalization? Are those systems not certain that this policy is no longer fruitful with this continuous flow of the mujahideen from east and west and this overwhelming awakening from the young people of Islam who no longer worry about their prisons and do not pay attention to their whips and do not take notice of their conferences and conspiracies?

Therefore, so that we can continue the march of compulsion to those arrogant superiors and we reach the end point that is "and there prevail justice and faith in Allah altogether and everywhere" [Koranic verse; al-Anfal 8:39], then it is our duty legally, rationally and realistically to remain steadfast on the path of jihad and we grasp on to it as we grasp "the most trustworthy handhold, that never breaks" [Koranic verse; al-Baqarah 2:256] and that we continue in fighting them and assaulting them by all that we own, enduring the burden of the battle, and taking into account the rewards of its wounds, its trouble, and its woes with God and that we double the efforts of incitement to the Islamic nation and to alarm and lower the matters of its enemies in its eyes and to send the hopes of victory and the good news of victory and empowerment in its hearts. As for cutting off the movement half way and retreating to the era of surrender to reality and submissive to the enemies of the Islamic nation like this forsaken document calls for, by alleging weakness and lack of support and the enormity of the duty that the Islamic nation is paying, all of this is nothing other than giving a break to those tyrants so that they can regain the respect of their thrones after they were shaken and to their countries and systems that became weakened and broken by the grace of God and then by the strikes of the mujahideen. How much those criminals wish – and they have tasted what we tasted of injuries, wounds, and pains – how much they wish that the jihad movement would lend them its ears or become pliant or give them sweet talk from their leaders, as God the Exalted says: "Their desire is that thou shouldst be pliant: so would they be pliant" [Koranic verse; al-Qalam 68:9]. And the Exalted says: "And their purpose was to tempt thee away from that which We had revealed unto thee, to substitute in our name something quite different; (in that case), behold! they would certainly have made thee (their) friend!" [Koranic verse; al-Isra 17:73]

Therefore, their hurt and their anger is due to the scorn of the mujahideen to them and to their countries and rules and their insistence on not accepting any point to agree with them is not less than their hurt and their discontent with the military strikes that shake them and remove the roots of their deviated fake civilizations. Those who are plotting day and night should know and they discharge their plots openly and secretly that the jihad today has become a jihad of a nation and the spirit of pride, a sense of honor, and the persistence that is flowing in its arteries like the flow of blood and that the days of deception are gone and the pages of degradation are erased and the policies of torture are discharged and that the mujahideen have known their way after they ignited the land with fire under the feet of your masters, so they will not turn back and they will not pay attention to the noise and nonsense; so carry out your cunning as you wish. For God is our Patron and there is no other patron for you. "Fain would they extinguish Allah's light with their mouths, but Allah will not allow but that His light should be perfected, even though the Unbelievers may detest (it). It is He Who hath sent His Messenger with guidance and the Religion of Truth, to proclaim it over all religion, even though the pagans may detest (it)" [Koranic verse; al-Tawbah 9:32-33]. Yet, we wonder: what are the motives behind the document? In the first place, what was the aim of those behind it when they released it at this time and under this circumstance?

In short, I say that the present battle of Islam against the forces of evil, manifested in the Crusaders and their hirelings, has lately taken three parallel approaches, each of which requires a detailed explanation:

The first approach was to intensify the military campaign against the mujahideen in the battlefields and to depend on the traitors and agents in doing so, as is the case in Iraq in the project known as the Awakening [Councils], which was joined by a group of people that sold its religion for the worldly gains of others. This approach can be confronted through steadfastness on the part of the mujahideen as well as their perseverance and endurance in the face of their enemy in the battlefields and by the removal of these national cancerous tumors, which stood side by side with the invaders and the occupiers and supported them. The only way with these malicious people is to defeat them and eradicate them in the same manner we treat their masters, by whom they seek protection and on whom they depend and whose plans they carry out. When they repent, stop their transgressions and keep their hands off, in reality and genuinely, and it will be then and then only that they will be told, "Go, you are free. You are not to condemned any more, and God is Most Merciful." Forgiveness corrupts the wicked as much as it mends the noble: "If you are kind to the noble you win him over, but if you are kind to the wicked he goes up against you" [Arabic proverb]; "placing dew in the place of the sword is as harmful as placing the sword in place of the dew".

The second approach was to flood the battlefields of jihad with deviated fatwas incriminating and forbidding joining them to get the youth of the nation to abstain from them. It was also to plant doubts about them and to return the matter to the hands of the people in authority, who are deep in betrayal, collaboration, immorality, and in fighting Islam, like the fatwa that was issued by the Mufti of al-Saud in which he included strange and ignorant arguments that were never heard before. He even claimed that the banners in these battlefields, which Muslims were forbidden to join, were doubtful. I do not understand! Could it be that the banner of Schwarzkopf, who led the armies of the liberation of Kuwait, was the one who was clear, simply because the ones who are carrying and calling for it are fighting to make God's word supreme? Such fatwas did not come by chance, even though in some cases, those who issued them were persuaded. They came as part of the campaign that aims at eradicating the mujahideen from within and blocking them from the outside, preventing their supplies, men, and funds from reaching them. Facing this approach will be by refuting these doubts, which are raised by the introducers of these deviated fatwas and confronting them with the clear truth, which they cannot reinterpret, fight, or manipulate. This is the duty of the scholars, upon whom God has placed the responsibility of delivering the truth so that they can be the heirs of our prophet, prayers and peace be upon him, in fact and not by claim. "And remember Allah took a covenant from the People of the Book, to make it known and clear to mankind, and not to hide it" [Koranic verse; al-Imran 3:187].

The third approach is to soften the worship of jihad and plant doubt about its foundation and principles on which it is founded, such as the obligation of jihad for the Muslims today and the need to drive out the occupiers and to disobey the apostate rulers and to bury the concept of allegiance and animosity and replace it with nationalism, patriotism and pre-Islamic prejudices. One of the means to carry out this task was what the evil minds in the headquarters of the Egyptian security apparatus have come up with in the document of their guidance. Therefore, I draw the attention of my mujahideen brothers in particular and the Muslims in general to the most important goals which this document is meant to achieve through twisted ways and hidden rumbles, which many of the people who will read it might miss. However, the detailed clearing of doubts will come in a detailed written document, with God's will and assistance.

The first of these goals is to distract the mujahideen, once in a while, by raising issues that cause turmoil in their ranks, and open a new front for them, leading them to mere intellectual conflicts and drowning them in the battle of responses instead of fighting in the battlegrounds which had exhausted the enemy militarily and economically, bankrupted it both intellectually and methodologically, and uncovered its crimes publicly. It exposed its civilization, on which it prides itself and for which it calls. Let the mujahideen beware of falling in this abyss or following what the enemies of Islam are claiming when they say that the battle between them and us is an intellectual battle for fear that this idea might be accepted in our minds, leading us to become inattentive to our weapons, while they are at the peak of their alertness and plotting, according to the words of the Almighty who said, "Taking all precaution, and bearing arms: the Unbelievers wish, if ye were negligent of your arms and your baggage, to assault you in a single rush" [Koranic verse; al-Nisa 4:102]. If an ideological conflict or an intellectual conflict as they call it, truly exists between us and our enemies, it should be from our point of view as Muslims and not from their point of view or their understanding. However, this should not be an alternative for fighting those who denied the Sharia of God or fought His followers and supported His enemies.

The second goal behind this document is to portray the mujahideen as a mob of looters, robbers, thieves, and followers of their whims, who have no values or ethics or discipline or religion and who are a group of arrogant illiterate [men] who are not led by any banner, disciplined by any concept, or controlled by any wisdom. This is a meaning that is repeatedly mentioned in the document of discouragement. Therefore, the author of the document presented these immoral meanings in form of advice, direction, guidance, and compassion, as if these evil characteristics, which he claimed against them, are true and do not need to be proven or refuted, and that all we need to do is busy ourselves with curing them. This blemished meaning of the mujahideen and their heroic work is adopted by the media in all its forms. It exerts all its efforts to confirm this meaning, and to support it, even at the least occasion, with lies and fabrications in order to separate the mujahideen from the Muslim nation, after they had been portrayed before her as a gang of immoral criminals, whose only cause and interests are to spill the blood, steal the money, and displace the secured. "It is a grievous thing that issues from their mouths as a saying what they say is nothing but falsehood!" [Koranic verses; al-Kahf 18:5]

The Almighty God has warned us of this type of the makhdhulin and the murjifin: "If they had come out with you, they would not have added to your (strength) but only (made for) disorder, hurrying to and fro in your midst and sowing sedition among you, and there would have been some among you who would have listened to them. But Allah knoweth well those who do wrong" [Koranic verses; al-Tawbah 9:47].

The third goal of the document of defeatism is the opposite of what I had mentioned earlier, for the authors of this document aimed at beautifying the picture of the infidel West, especially America, in the eyes of people. They wanted to polish their image in the eyes of the people and portray them as just, gentle, and kind people who shelter the oppressed, and thus we had to deal with them on equal footing and follow their good example and virtuous ethics to strengthen the concept of subordination to and fascination with their civilization. The document of humiliation conceals the daily crimes that are committed by these butchers and overlooks those atrocities, the likes of which the human race has not seen in mass killings, eradication of people, continuous humiliation, rationed embezzlement, and others. It is a document that gives an ugly picture of the mujahideen and an embellished pleasant picture of those infidels.

The fourth goal of the document of defeatism is to hypnotize the Muslim nation and to plant in it the spirit of the acceptance of reality and to let the meaning of submission sink deep in their hearts and ensure that it becomes a deep-rooted incurable, unavoidable condition [of submission]. Thus it will be a waste of time to think of changing this reality or eradicating it. The duty will be to recognize it and deal with it as an existing reality, which one cannot deny or from which one cannot turn away. It will be useless to try to establish another. This concept, in its entirety, is the first step toward calling for global peaceful coexistence and strengthening the concept of nationalism instead of the allegiance and animosity concept. All of this is to be achieved through defeatism and spreading false reports and discouragement, which should be confronted with every strength and must be responded to very firmly. We should refuse to go back to slavery, defeat, and humiliation after having taken these steps and crossed these distances on the path of empowering this true religion.

There are other major and dangerous goals, which time does not allow for, which will be discussed, God willing, in the detailed response to the doubts which were raised in the document. Oh relentless and enduring mujahideen in Afghanistan, Iraq, Algeria, Somalia, Palestine, Chechnya, Egypt and every other Muslim spot in the world: reject these calls for weakness and submission and follow the advice of your prophet, prayers and peace be upon him, "You should know that victory comes with endurance, and the relief comes through distress, and along with difficulty comes ease" [Hadith]. Hearing the news is not like seeing it. For whoever tasted jihad, witnessed its glory, felt its value and saw its effect cannot be impacted or shaken by aimless doubts or weak suspicions, no matter how embellished. You must know, brothers of jihad, that the release of such documents from the prisons of criminality and vileness and attributing them to people who were prominent figures in the jihad field confirms the obligation of jihad and increases your responsibility because these prisoners under the weight of convincing criminality and continuous humiliation were forced to write these wrong and misleading words. Take upon yourselves the responsibility of freeing them, with all you can, and exert every effort to lift oppression, constraints, and imprisonment off of them instead of busying yourselves with slandering and belittling them. Remain steadfast on the path of jihad, and you will increase in power and glory, while your enemy will suffer more defeat. Tell the agencies of deception and vileness: "I Am not a Deceiver, nor Will I Allow Someone to Deceive Me! [This was originally said by Muslim commander Umar Ibn-al-Khattab.] Your only medicine is the sword."

"Fight the unbelievers who gird you about, and let them find firmness in you: and know that Allah is with those who fear Him" [Koranic verse; al-Tawbah 9:123]. Our last supplication is to praise God, Lord of all creation.

The Moderation of Islam and the Moderation of Defeat

Release Date: 22 May 2008
Production Date: Unknown
Type: Video statement
Version: Translation

In the name of God, the Merciful, the Compassionate

Praise be to God, and prayers and peace be upon the messenger of God and upon all his family, his companions, and all those who followed him thereafter.

O nation of Islam, east and west of the globe,

May peace and the mercy and blessings of God be with you.

When we talk about any of the issues of Islam, we must feel in the depth of our hearts that this religion, the matters of which we are talking about, is the religion of Almighty God. It is a feeling that has its direct impact on how we decide on the issues, discuss them, and inquire into them. Islam – all of Islam – is not a worldly theory that is subject to free and unrestricted investigation, without control or rein. Instead, it is "A Book revealed unto thee -- So let thy heart be oppressed no more by any difficulty on that account – that with it thou might warn (the erring) and teach (the Believers). Follow (O men!) the revelation given unto you from your Lord, and follow not, as friends or protectors, other than Him. Little it is ye remember of admonition." [Koranic verse; al-A'raf 7:2]

Since the religion is that of the Almighty God, there is no place for tampering with it to suit one's likings; for engaging in ideas, opinions, mind games, and analysis; for concerning oneself with pleasing people; for submitting to the calls to keep up with the modern age; or for singing to the desires of the nations. The religion of God leads and will not be led. It overpowers the people and will not be overpowered by them. It rules the nations and does not get ruled by them. It restricts the desires, while desires do not restrict it, and it controls the concerns of the age, while they do not control it. It completely dominates life, but life does not dominate it.

Therefore, those who want to save the earth from corruption and get it out of the darkness of temptations by depending on their personal convictions, their opinions, and their inclinations, under attractive slogans that are in fact empty and far from the path of the truth and guidance, they will gain nothing from their quest except negative consequences and mental disorder. As the Almighty says: "If the Truth had been in accord with their desires, truly the heavens and the earth, and all beings therein would have been in confusion and corruption! Nay, We have sent them their admonition, but they turn away from their admonition." [Koranic verse; al-Mu'minun 23:71]

Since we are in an era in which the god of desire has reached the highest rank ever, has exploited large armies to support and strengthen him, and has opened to the lofty edifice of Islam several fronts to weaken it and uproot it, we have a serious need for a firm and sincere stand in the face of anyone who wants to be among the soldiers of the god of desire, who has appeared to us in various shapes and colors, and to whom many have bowed in praise and pumped up his divinity, knowingly or unknowingly, to move with awareness of our situation to defend our principals, our religion, and our beliefs, and to protect them from any watering down or corruption. "Then We put thee on the (right) Way of Religion: so follow thou that (Way), and follow not the desires of those who know not. They will be of no use to thee in the sight of Allah. it is only Wrong-doers (that stand as) protectors, one to another: but Allah is the Protector of the Righteous. These are clear evidences to men and a Guidance and Mercy to those of assured Faith." [Koranic verse; al-Jathiyah 45:18-20]

O Muslim nation

There is a cooperative company managed by devils among men and jinns that has its men and capabilities, as well as its means and its establishments, its expenses and efforts, its plans and its programs. This company works assiduously to divert people from their religion; plants doubts in their minds about their fundamental beliefs; and encourages anyone who wants to engage in it in his own way in the name of ijtihad, intellect, debate, enlightenment, analysis, researching reality, openness and balance to the end of the well-known list, without having an advocate to defend him, a maximum limit to stop at, or rules and principles to avoid. Instead, everybody has unlimited room to condemn – without shame or fear – the clear truth. He very audaciously alters the words and attributes to the law of God what is known to the elders in the desert as far from it. He rejects what was known and accepted by the first and the last. You even see him discrediting and mocking them and ridiculing their perceptions. He covers his hallucinations with titles and praises, which he desperately but unsuccessfully uses to cover up their foolishness and abjectness.

The prophet, prayers and peace be upon him, told us the truth about this evil partnership, and that it is found in every age to carry out its mission and to tempt those who pay attention to it or listen to its calls. [He warned us] so that we can be careful and attentive not to be deceived by its propaganda and its fancy words. Abdallah Ibn-Mas'ud, may God be pleased with him, reported that God's messenger (peace be upon him) once drew a line in the dust with his hand and said, "This is the straight path of God." Then he drew a series of lines to the right and the left and said: "Each of these paths has a devil at its head, inviting people to it." He then recited (Koran 6:153): "Verily, this is My way, leading straight. Follow it. Follow not (other) paths. They will scatter you about from His (great) path. Thus doth He command you, that ye may be righteous." [hadith]

Almighty God also said: "Likewise did We make for every Messenger an enemy – evil ones among men and jinns, inspiring each other with flowery discourses by way of deception. If thy Lord had so planned, they would not have done it. So leave them and their inventions alone." [Koranic verse; al-An'am 6:112]

The Almighty also said, "O ye Children of Adam! Let not Satan seduce you in the same manner as He got your parents out of the Garden, stripping them of their raiment to expose their shame. For he and his tribe watch you from a position where ye cannot see them. We made the evil ones friends (only) to those without faith." [Koranic verse; al-A'raf 7:27]

This was God's commandment to His prophet, prayers and peace be upon him, to whomever wants to be on the path, following the way of truth, without concerning himself with those who disagree with him. He holds on to the right path without worrying about who denies him and who disowns him. He holds on to the rope of God without paying heed to any ridicule or defamation. Instead, he concerns himself with pleasing his God, even if the whole world resents him. He is someone who takes pride in his religion, even if it is discredited by others. He conveys the truth in its purity and its clarity, even if it is disliked by those who have deviant minds and sick hearts. His ideal person is someone who does not speak out of his own desire and who does not discriminate in proclaiming the message of God, according to the words of the Almighty: "O Messenger, proclaim the (message) which hath been sent to thee from thy Lord. If thou didst not, thou wouldst not have fulfilled and proclaimed His mission. And Allah will defend thee from men (who mean mischief), for Allah guides not those who reject Faith." [Koranic verse; al-Ma'idah 5:67] He also said: "Therefore expound openly what thou art commanded, and turn away from those who join false gods with Allah." [Koranic verse; al-Hijr 15:94]

My beloved nation,

Until recently the leaders of the modern Crusader campaign were raising the banners of fighting terrorism and hunting down al-Qaeda, its leaders and members – as they claim. The wise and intellectual of the Islamic nation's clerics and the mujahideen leaders said at the first gusts of wind of this war that this is a clear infidel Crusader war against Islam and Muslims. The efforts and steps of its possessors will not stop at a limit, and they will not be satisfied, as they allege, at the perseverance to eliminate the mujahideen group and one concession will not pass until they demand another. In fact, their campaign has not left any of the principles of Islam or its fundamentals alone. It turned right and left to uproot and demolish Islam in order to transform the Islamic nation – the entire Islamic nation – from clarity to blindness, from certainty to doubt and from belief to disbelief. That is their desire and intention, no matter what color they take or change. The Koran will expose them, just as it exposed their forefathers. And just as it exposed those, it will expose them. God the Exalted says: "Never will the Jews or the Christians be satisfied with thee unless thou follow their form of religion." [Koranic verse; al-Baqarah 2:120]

It is a war that targets all of the strongholds of Islam. It invades homelands and penetrates minds and thoughts. It dares to shed blood exactly as it dares to destroy beliefs and tamper with the sacred.

The chiefs and clever men of this crusade were able to plant for themselves people from the sons of this religion and the heart of the Muslim countries who are in charge of marketing many of their ideas, promoting their theories, transmitting their terminologies, repeating their phrases, and trying to convince Muslims with them, or at least killing the feeling of its ugliness and gravity, so that after a while it will become something pleasant, an acceptable idea, and a considered view.

And because they knew that the key to their success in this plan of theirs is to turn the people away from jihad and mujahideen and to eliminate them militarily and intellectually. Whenever any obscure person merely speaks and mumbles words that expose the mujahideen, their media will rush to make him prominent and famous, and they will conduct continuous discussions and successive interviews to guide the worshipers of God to this unique idea that burst out of a madman's mind and that barely makes sense.

Regrettably, we saw many who are related to the leaders of the Islamic movements or proselytizers or intellectuals who took it seriously and continued to hold one conference after another, one interview after another, and one symposium after another. They travel from one country to another to support misleading concepts that target Islam at its base. Following them only means destroying the foundation of Islam little by little. They attribute their straying and deviation to the religion of God. Thus, they added a sin to their sins. The word of God applies to them, when the Exalted said, "There is among them a section who distort the Book with their tongues. (As they read) you would think it is a part of the Book, but it is no part of the Book. They say, 'That is from Allah,' but it is not from Allah. It is they who tell a lie against Allah, and (well) they know it!'" [Koranic verse; al-Imran 3:78]

The greatest disguise they use to hide their crookedness and to market their misguidance is their following of the middle path of moderation and balance. They have fashioned meanings for these words that they accepted for themselves, which they carved from their own minds, based on fabrication and suitableness and using the technique of liquefaction and distortion to delight the West with what pleases it and calms its fury, even if this means blasting the religion of God in its entirety.

Then, what is this "Wasatiyyah" that those people are calling for and buzzing around it day and night? What is the justly balanced "Wasatiyyah" that the religion of God, the Great and Almighty devised and wanted for us and praised the nation of His prophet by it when He says: "Thus, have We made of you an Ummat justly balanced, that ye might be witnesses over the nations, and the Messenger a witness over yourselves"? Koranic verse; al-Baqarah 2:143]

Nation of Islam

The determination of the concepts of Islamic law that we formulate for it and the terms that improve and beautify it must depend on the book of God and the methodology of His prophet, prayers and peace be upon him, because minds vary, thoughts are contradicted, the scales of matters are confused, and desires and wants creep into this or that meaning so that it tampers it and taints its glory. Then there must be a stable, deep-rooted, and fixed reference that contains no wrong from any direction. If it says that, it is the words of righteousness. If it rules, then the verdict is just. If it leads, then it guides to the right path and by nothing but the book of God, as the Great and Almighty says: "If the Truth had been in accord with their desires, truly the heavens and the earth, and all beings therein would have been in confusion and corruption! Nay, We have sent them their admonition, but they turn away from their admonition." [Koranic verse; al-Mu'minun 23:71]

Moderation is a word people accepted, found agreeable, and perhaps had agreed on, praising it for its sense of justice, balance, and reasonability. But many who have assigned themselves to call for it have emptied it from its lofty Islamic law and poured into it whatever they desire and their tendencies to accept oblique sense and concepts without the authority from God. Then they present it to the people and tell them that this is moderation, that they must follow it and reject those who oppose it or who did not join it.

The right concept of moderation is abiding completely to the religion of God, the Exalted, that He wanted for all the people, even if some did not like it, and to strive to spread it amongst them without deviation or falsification or maneuvers, and to present it to them in a frank clear presentation, without playing with its rules, destroying its fundamentals, changing of its laws, or concealing its facts, and without the embarrassment of deciding its matter, then let it be accepted by those who accept it and rejected by those who reject it. "Therefore do thou give admonition, for thou art one to admonish. Thou art not one to manage (men's) affairs." [Koranic verse; al-Ghashiyah 88:21-22]

We were not ordered to kill ourselves, and we fret about it for the sake that the people would be repelled and go astray from it, but we have to transmit the open truth and hold strongly to the obvious right path.

God the Exalted says to His prophet, prayers and peace be upon him: "Thou wouldst only, perchance, fret thyself to death, following after them, in grief." [Koranic verse; al-Kahf 18:6] The Exalted says: "It may be thou frettest thy soul with grief, that they do not become Believers. If (such) were Our Will, We could send down to them from the sky a Sign, to which they would bend their necks in humility." [Koranic verse; al-Sh'ara 26 3-4]

For the attribute of belief and the propagation of virtue prevention of vice, this nation held the best status among all nations: "Ye are the best of peoples, evolved for mankind, enjoining what is right, forbidding what is wrong, and believing in Allah." [Koranic verse; al-Imran 3:110] So it deserved to become a justly balanced nation as a witness over the people, as the Exalted says: "Thus, have We made of you an Ummat justly balanced, that ye might be witnesses over the nations, and the Messenger a witness over yourselves." [Koranic verse; al-Baqarah 2:143]

Then the duty of the nation of Islam is not to go along with the nonbelieving nations, to sweet-talk them, to work to please them, to try to find common ground with them, to search for the basis of coexistence that brings them together, or to expend efforts and waste one's whole life trying to convince them to accept reality and surrender to it. We were not created for this, nor were we ordered by this (God created us and brought us so that we can transform whomever He wants from worshiping people to worshiping God, from hopelessness to hopefulness, and from the injustice of other religions to the justice of Islam. He sent us with his religion to His creation for us to invite them into it. For those who accept this from us, we accept this from them. For those who refuse it, we will fight them to the end, when we go to God's promise in heaven for those who fought the one who refused, and victory is for those who are still alive). This is how "Rab'i Ibn-Amir," may God be pleased by him, summarized the duty of the Islamic nation. This is the true "Wasatiyyah," as understood by the companions, may God be pleased by them, and called for it, away from the cold lies, misleading complications, and floating ideas.

No one has the right to choose from the religion of God what he likes and desires, nor is he able to bend the rules of God for those he loves and likes or present the religion of God in the manner that he likes and wishes. The Exalted says, "Say thou: This is my way: I do invite unto Allah, on evidence clear as the seeing with one's eyes – I and whoever follows me. Glory to Allah, and never will I join gods with Allah." [Koranic verse: Yusuf 12:108] Then it is a call to the path of God – which means the entire religion – and is not an invitation to abstract opinion outcomes or the preferred new-fashioned ideas. God the Most High says: "O ye who believe! Enter into Islam wholeheartedly and follow not the footsteps of the evil one, for he is to you an avowed enemy." [Koranic verse; al-Baqarah 2:208]

For Islam, in its entirety, is a religion of justice and the religion of justly balance and of value. The people will not succeed and will not find (true Wasatiyyah) by any means other than taking the religion as is and spreading it as is. It is a divine religion with no need for our changes to make it moderate and make it right. Its rules are not unjust rules that we need to amend, and its path is not a path of excess and immoderation for us to balance. Its laws are not crooked laws for us to set straight. "Allah commands justice, doing good, and liberality to kith and kin, and He forbids all shameful deeds and injustice and rebellion. He instructs you, that ye may receive admonition." [Koranic verse; al-Nahl 16:80] And the Most High and Glorious says: "Allah doth command you to render back your Trusts to those to whom they are due. And when ye judge between man and man, that ye judge with justice. Verily how excellent is the teaching which He giveth you! For Allah is He Who heareth and seeth all things." [Koranic verse; al-Nisa 4: 58]

Moderation is not legitimate evidence that exists by itself so as to have it dominate the religious laws of Islam, overpowering its beliefs and concepts and constraining its grounds and branches. Rather, it is an attribute attached to the religion of the Almighty God and goes side by side with every one of its rules. For wherever the rule of God exists, which He revealed in His Book [Koran] or legislated through the sayings of His prophet, prayers and peace of God be upon him, whether in small or big issues, it is a rule of moderation, justice, and forgiving, and it has no confusion or adversity, even though some people find it weighty and some hearts find it repulsive.

The infirmity does not lie in God's rules in order for us to try to correct and adjust it in order to coincide with those souls and hearts, but the cohesive infirmity is in those souls and hearts that need to be drawn out of their madness and their devilish insinuations in order to have them perceive the just moderation, as it is with its transparency and magnificence, not in accordance the way that they desire or love. By means of this alone can one avail oneself of the truth and know the mercy of justice and the amplitude of moderation. Otherwise, they will continue to overturn in its temptations and suffer of its wretchedness, even if they thought that they are doing something good. "Is then one who is on a clear (Path) from his Lord, no better than one to whom the evil of his conduct seems pleasing, and such as follow their own lusts?" [Koranic verse; Mohammed 47:14].

We now behold many of those who have raised the slogan of moderation and have become, allegedly, saturated with it and have made it an excuse for violating the undisputed basics principles of Islam without caring. Every time he encounters opposition and disapproval, he would accuse them of exaggeration, extremism, and lack of openness to the state of things as they are. By doing so, they open the door to those who are great nonbelievers and atheists to enter upon Islam from the same entrance – that of alleged moderation – thus spreading evil, not leaving an irregularity or an abnormality, but that they would run after its with their thoughts, hound it with their mockery and blast it with the axe of their moderation and moderateness. And anyone who tries to assist them with their nonbelief, they would accuse him of extremism, transgression, and rigidity and then they will go on to destroy Islam as a whole.

So in the name of "Wasatiyyah" and moderation, the domes of polytheism were entered not for the sake of conquering them and establishing therein the banner of monotheism, but to become receptive toward them and to strengthen their foundation, assert their legitimacy, and collaborate with their heads in making their laws. So if the truthful people declared a word of truth that they were commanded to declare and recited the splendid wonders of God to others, their hearts would be appalled and they would accuse them of exaggeration, extremism, and a lack of political understanding. By God, their policy and moderation will not do them any good in God. "The Day that (all) things secret will be tested, (Man) will have no power and no helper." [Koranic verse al-Tariq; 86:9-10]

So if it were to be told to those who are saturated with moderation: "It is not fitting for a Believer, man or woman, when a matter has been decided by Allah and His Messenger to have any option about their decision. If anyone disobeys Allah and His Messenger, he is indeed on a clearly wrong Path," [Koranic verse al-Ahzab; 33:36], they would say: You are too stern.

"And if they are asked if they are not rebuked by the Almighty's saying: 'They take their priests and their anchorites to be their lords in derogation of Allah, and (they take as their Lord) Christ the son of Mary; yet they were commanded to worship but One Allah. there is no god but He. Praise and glory to Him: (Far is He) from having the partners they associate (with Him).'" [Koranic verse; al-Tawba 9:31], they would say that you are close-minded deviants.

And if they were told: "Have they partners (in godhead), who have established for them some religion without the permission of Allah." [Koranic verse; al-Shura verse 21], they would say: you are stern and austere.

And if they were told: "Were thou to follow the common run of those on earth, they will lead thee away from the way of Allah. They follow nothing but conjecture. They do nothing but lie" [Koranic verse; al-An'am; 6:116], they would say we are in an era of democracy and sovereignty of the people.

And in the name of moderation and moderateness, the allegiance and disavowal creed has been destroyed, and its bond, the strongest bond of faith, has been ripped.

Its fortified fortresses have been stormed under the adorned slogans and fabricated calls, and we began to hear of civilizations that keep peace with each other and do not clash, that converse and do not quarrel; and religions that stand together and do not dispute each other and come closer and do not fight. Some of those who belong to the call stood up to spread these calls of nonbelief, the facilitation of their course in the Muslim lands, and the distortion of the discourse from its rightful meaning – while in their hearts they know – that they are liars and untruthful. Woe to those who fabricate lies about God. You ward off God and you augment and decrease His religion as you wish, so how can you rule [among men]?

Such that I once heard one of those who spread such calls talking with pride that a number of priests and monks contacted him to inform him that they were satisfied with his call and tolerance, and that if the concepts of true Islam were as he presented them and publicized then there would be no differences between them. So good for you these glad tidings and good for you this testimony and I said to myself, yes by God for if Islam is what you preach then there is no difference between you and them.

As for true Islam, the transparent and pure creed and the immaculate monotheism it is utterly impossible for it to come close to a religion that claims that "Jesus is the son of God"and that 'God is a Trinity'. Even the Jews, may God destroy them, said, because of their many contradictions with the prophet, prayers and peace of God be upon him, in small and big matters: this man does not want to leave anything except to contradict us on.

So the moderation that we are calling for says: "Thou wilt not find any people who believe in Allah and the Last Day, loving those who resist Allah and His Messenger, even though they were their fathers or their sons, or their brothers, or their kindred." [Koranic verse; al-Mujadilah 58:22] Indeed, if you consider this exaggeration.

And the moderation that we follow is based on: "O ye who believe! take not for protectors your fathers and your brothers if they love infidelity above Faith: if any of you do so, they do wrong." [Koranic verse; al-Tawbah 9:23], even if you vilify us and accuse us regarding our patriotism.

The moderation, the foundation of which we are establishing is calling [upon the people] everywhere, "O ye who believe! take not for friends and protectors those who take your religion for a mockery or sport,- whether among those who received the Scripture before you, or among those who reject Faith; but fear ye Allah, if ye have faith (indeed)." [Koranic verse al-Ma'idah; 5:57], even if you are meet this with disgust and fury.

The moderation that we defend and fight for has as its pioneer and its leader about whom God said, "There is for you an excellent example (to follow) in Abraham and those with him, when they said to their people: 'We are clear of you and of whatever ye worship besides Allah. we have rejected you, and there has arisen, between us and you, enmity and hatred for ever,- unless ye believe in Allah and Him alone'" [Koranic verse al-Mumtahinah; 60:4], even if you say that this is a call for hatred, fanaticism and fighting against peace.

Our moderation that we hold tight to and do not deviate from not even by a fingertip is what we find in God Almighty's saying: "O ye who believe! take not the Jews and the Christians for your friends and protectors: They are but friends and protectors to each other. And he amongst you that turns to them (for friendship) is of them. Verily Allah guideth not a people unjust" [Koranic verse; al-Ma'idah 5:51], even if you consider that to be extremism or rigidity.

At this point I say: O you scholars of verity, who declare the truth in the Arabian Peninsula, there stands the protector of monotheism, as claimed by the scholars of flattery and cajolery, carrying the banner of brotherhood between religions and hallucinating about what he does not know, thinking that he has found the wisdom that others have labored in order to find and remove the cord of wars and cut off the causes of hatred between the religions and peoples. This is your day: the hurling of accusations has increased; the crows of untruthfulness have croaked loudly, and the sulky face of infidelity is exposed, and the scholars of flattery have supported the darkness of misguidance, and plunged into the depth of books to bring out the most precise of suspicions and eliminate the clear and impermeable verses. They exhausted themselves in order to make up excuses for that foolish [man] and his party who never thought it would happen, nor did they dream of it at any time day or night.

It is by God, a time of victory for he who wants to be a master of martyrs. He stands up in the face of that apostate insane [man] and responds to his nonsense with pure truth that has no ambiguity or subtleness. His blood and body would later serve as offering that would quench the withered tree of Islam in the Arabian Peninsula, as the land of Pakistan was quenched with the blood of the people of truth and verity, such as martyr Abdul Rashid Ghazi and his faction, may God have mercy upon them.

And by God, if you do not stand today as heroes in the face of this frivolous tyrant, and [instead] clear the way for the scholars of misguidance, the preachers of the clergymen's prose, who justify and legitimize his deviations and foolishness, then the day will come when you will hear the bells of churches ringing in the heart of the Arabian Peninsula. The case in the State of Qatar is not far-off from you, and you will hear the news soon.

So there is no moderation or rapprochement between us and between the people of infidel factions. Since when have light and darkness met in one spot?

Almighty God said: "So if they believe as ye believe, they are indeed on the right path; but if they turn back, it is they who are in schism. But Allah will suffice thee as against them, and He is the All-Hearing, the All-Knowing". [Koranic verse; al-Baqarah 2:137].

This is our prophet, prayers and peace be upon him, who differentiated between people, so those attempting to combine them without the right path are doing so in vain.

In the name of moderation, the evil spreader tampered with the concept of jihad by targeting it with their pens and tongues. They stained its existence, tarnished its glamour, and emptied its rulings in the mould of their defeat, cowardice, and weakness to come up with a distorted form that was unknown to the ancestors and unacceptable to the faithful, but it is sufficient that the civilized infidels of the West are convinced.

Previously, the attack, despite its heinousness and hideousness, was limited to preemptive jihad, which the defeated were perplexed in directing it, and lower their heads with shame when a mention of it is made, and feel embarrassed if their enemies confronted them with its reality.

Today, defensive jihad is facing the same attempts at distortion and discouragement, in addition to the efforts to water it down and alter it. Jihad became resistance, and resistance was divided to honorable and dishonorable resistance. Its legitimacy was defined not by the true book, the sound Sunnah, or a firm consensus, but by the approval of all the divine religions and the international norms to give this right.

They suppressed the Islamic characteristics of even this defensive jihad, and replaced them with patriotic and nationalistic slogans. The sublime meanings of jihad dissolved in the confused political concepts. Its strict rulings dissolved in the name of indecisive moderation. Its noble truth was lost amid the turmoil of rationality, serenity, and interest.

Misrepresentation and defeatism had reached a stage in which some have said that our fight against the Jewish occupiers and the criminal Christians is not a religious fight, but only a conflict over the lands they occupied and the homes they seized unlawfully. This is the penalty for those who stray from the truth and follow their accepted notions and their desires. They move from one deception to another, from one ambiguity to another, and from one deviation to another. God said: "Then let those beware who withstand the Messenger's order, lest some trial befall them or a grievous penalty be inflicted on them." [Koranic verse; al-Nur 24:63]

Let those and others who falsely hide under the garment of moderation know that Islam is the religion of the sword. We say it, and we are not ashamed of it. We do not avoid saying it loudly. Our prophet is the smiling fighter, the prophet of guidance and the epic. The sword and monotheism are never separated. There is no acknowledgment of monotheism without the sword and power, and there is no meaning of the sword if it is not for the sake of monotheism. Almighty God said: "And fight them on until there is no more tumult or oppression, and there prevail justice andfaith in Allah altogether and everywhere. But if they cease, verily Allah doth see all that they do." [Koranic verse; al-Anfal 8:39]

And the prophet, prayers and peace be upon him, said: "I was sent before the hour with the sword, until Allah the Exalted is worshipped alone with no partners." [hadith]

The prophet, prayers and peace be upon him, said: "I have been ordered to fight against people until they testify that there is no god but Allah and that Mohammed is the messenger of Allah and until they perform the prayers and pay the zakat, and if they do so they will have gained protection from me for their lives and property, unless they do acts that are punishable in accordance with Islam, and their reckoning will be with Allah the Almighty." [hadith]

We do not wait for legitimacy to be granted to our offensive or defensive jihad from any religion besides Islam, for a law by a global organization, international legitimacy, regimes that have colluded with worldly customs. All of these are idols that must be demolished and tyrants whose elimination must be pursued. It is the first that must be considered as an infidel and must be denounced. If its people glorify it and spend their precious money on it, and waste their youth to spread it, to us those who promote it do not represent a wing of an insect and more despised than rats. God said: "Verily ye, unbelievers, and the false gods that ye worship besides Allah, are but fuel for Hell! to it will ye surely come! If these had been gods, they would not have got there! but each one will abide therein." [Koranic verses; al-Anbiyaa 21:88-89]

It is the religion of Islam and not the United Nations or the international norms that said to us: "Fight those who believe not in Allah, nor the Last Day, nor hold that forbidden which hath been forbidden by Allah and His Messenger, nor acknowledge the religion of Truth, even if they are of the People of the Book, until they pay the Jizyah with willing submission, and feel themselves subdued." [Koranic verse; al-Tawbah 9:29]

It did not say to us, do not fight them if they do not occupy, invade, or dominate your land. The matter does not need the philosophy of the defeated moderation, or the fabrication of the false rational minds, or the illusion and the misguidance of the enemies, to inform them of the opposite of what we were commanded. Either they believe in God or they submit to the ruling of Islam. If they refuse, we seek God's support and fight them. It is a clear matter which is found in every Fiqh book, but the eyesight of the defeated could not see it when they their eyesight has become blinded. "But say not – for any false thing that your tongues may put forth – this is lawful and this is forbidden, so as to ascribe false things to Allah. For those who ascribe false things to Allah, will never prosper." [Koranic verse; al-Nahl 16:116]

In the name of moderation and self-control, the cunning sought to abolish and erase many of the sharia expressions and words that irritate the infidel West and prevent rapprochement and understanding. They used the word "the other," instead of the word "the infidel," and they replaced the word "atheist" by the word "non-Muslim," and they portrayed the religion of Christianity and Judaism as "Divine religions." Even some of them exceeded the limit and called the Jews and Christians as believers.

Their tongues got used to such words, their pens documented, and their sites and discussions were filled with them. It became hard for the people of moderation to utter what was stated in God's Book and the Sunnah of His prophet, prayers and peace be upon him, until their situation became like those whom God said about, [God said:] "When Our Clear Signs are rehearsed to them, thou wilt notice a denial on the faces of the Unbelievers! they nearly attack with violence those who rehearse Our Signs to them." [Koranic verse; al-Hajj 22:72]

As for the moderation of Islam, it is what has divided the people into two groups that have no third, "Some He hath guided: Others have by their choice deserved the loss of their way." [Koranic verse; al-A'raf 7:30]

Either he is from the believing people or from the non-believing people, Almighty God said: "It is He Who has created you. And of you are some that are Unbelievers, and some that are Believers: and Allah sees well all that ye do." [Koranic verse; al-Taghabun 64:2]

These distorted calls that slowly and in secrecy infiltrate religious issues by misrepresentation and elusiveness and many underestimate its effect, but some of them consider it the utmost victory and a ruling above rules. I say, if these calls are not resolutely, boldly and openly confronted by the nation's truthful scholars and the zealous proselytizers, its consequence will be the creation of a new religion. I see its features have begun to be formed. It is a new religion in its terminology, new in its perception, new in its measures and principles, new in its creed and its branches, and even new in its worship and transactions and new in its reception sources. Then we will be told that this is moderate Islam, prudent Islam, Islam of the 21st century, Islam of openness, brotherhood, and peace, Islam of moderation and reason. It is only the Islam of the Rand Corporation and the enemies. The Islam that the infidel proselytizers are after by their modern Crusade campaign, this will never happen. Die with your anger and sorrow, God's religion is protected. A person can only destroy himself.

God said: "The Unbelievers spend their wealth to hinder man from the path of Allah, and so will they continue to spend. But in the end they will have only regrets and sighs; at length they will be overcome. And the Unbelievers will be gathered together to Hell, in order that Allah may separate the impure from the pure, put the impure, one on another, heap them together, and cast them into Hell. They will be the ones to have lost. Say to the Unbelievers, if now they desist from Unbelief, their past would be forgiven them; but if they persist, the punishment of those before them is already a matter of warning for them." [Koranic verses; al-Anfal 8:36 to 38]

This is the religion of Islam by its true moderation and complete justice and its clear call, not the misrepresentation of the defeated souls and sick hearts and misguided minds, even if thousands applauded them. God said: "The truth is from your Lord: Let him who will believe, and let him who will, reject it." [Koranic verse; al-Kahf 18:29]

Our last supplication, praise to God, Lord of all creation.

The Eid al-Fitr Speech for 1429H

Release Date: 28 Oct. 2008
Production Date: Unknown
Type: Video statement
Version: Translation

All that they said was: "Our Lord! Forgive us our sins and anything We may have done that transgressed our duty: Establish our feet firmly, and help us against those that resist Faith" [Koranic verse; al-Imran 3:147].

Praise be to God, whom we praise and from whom we ask for aid and seek forgiveness. We seek the exorcism from the evils of our souls and from the sins of our actions. He who is called to righteousness by God will never be led astray; and he who leads others astray can never be led to the path of righteousness. I witness that there is only one God, and He has no partners. I witness that Mohammed is His slave and His prophet. He is the best of His creations and His people. [Mohammed] was sent by God with the message of proselytization and the true religion to reveal it to all other religions, even if that was detested by the non-believers. May the prayers of God be upon him and upon his family and companions and upon those who accept his proselytization and follow his path of Sunnah until Judgment Day.

Praise be to God who has led us on the path to this point. We would not have followed the righteous path if it was not for God. Praise be to God who made us Muslims. Praise be to God who made us of the people who say "There is no god but God and Mohammed is the prophet of God." Praise be to God who made us of the people who know the truth of these words. Praise be to God who chose us from among his creations to make us His worshippers who forsake everything for Him, conduct jihad [for him], and remain steadfast [for his sake]. These are all blessings of God Almighty upon us, and they are the most glorious, wonderful, and magnificent. "But if ye count the favors of Allah, never will ye be able to number them" [Koranic verse; Ibrahim 14:34]. Oh you who have accepted God as your god, Islam as your religion, and with Mohammed, prayers and peace of God be upon him, as the prophet and messenger. Remember the blessings of God

Almighty upon you. Increase mention of [these blessings] so that you may increase your thankfulness of them. Thankfulness is the provision of blessings and only through being thankful does a man receive more blessings from God Almighty. "And remember! Your Lord caused to be declared (publicly): "If ye are grateful, I will add more (favors) unto you" [Koranic verse; Ibrahim 14:07].

Oh brothers, this is the journey of life. Coming and going; presence and absence. Yesterday, we were awaiting the month of Ramadan, and today we bid farewell to one of the stations of life which has been fulfilled, ended, and gone. Blessed be he who fulfills it with the obedience of God Almighty. Blessed be he who fasted during the month with faith and in fear of God. Blessed be he who has prayed during the month with faith and in fear of God. Blessed be he who has filled his time during the month with mentions of God Almighty, the recitation of His Book, praising Him, repentance unto God, and seeking forgiveness from God.

Woe unto those who have wasted these precious days which are the most precious days of life. Woe unto those who wasted them in disobedience of God Almighty. Shame on those who spent the time hibernating in front of the screens of waste and degeneration; shame upon those who spent the time in the markets of clamor, in the worlds of pleasure, and in the worlds of waste. Oh brothers, we have bid farewell to the month of Ramadan and the days we spent will never come back to us, but it is according to that with which we have filled these days, whether obedience or disobedience, honesty and goodness, or evil and the violation of the command of God Almighty. We exorcise ourselves from disobeying Him. This station [the month of Ramadan], was granted by God Almighty to be days with which the believer becomes closer to the people of righteousness and the people of goodness and good deeds. The Almighty said: "O ye who believe! Fasting is prescribed to you as it was prescribed to those before you, that ye may (learn) self-restraint" [Koranic verse; al-Baqarah 2:183]. "That you may learn self-restraint" is this amazing gift and blessing which God Almighty desired his worshipping believers to attain. It is the gift of devotion, and the blessing of devotion is from God Almighty. Is there anything more glorious, magnificent, bigger, or more superior than for a human being to be devoted to God Almighty? Devotion is the key to every goodness in the world and in the after-life. He who fears God will be given a passage to paradise and will be blessed from unexpected sources. He who fears God will be given ease in matters. He who fears God will be forgiven his sins, and his rewards will be multiplied. Therefore, this month was decreed by God as an opportunity for His worshippers to be elevated to the highest levels and to the most superior levels which God Almighty desires for them.

The believer, when he stops and thinks about his circumstances and looks at his deeds and reflects on how he has spent these days, should praise God if he sees that he has done good deeds, but if he finds otherwise, he should only blame himself. Yes, it is an opportunity of which the human being should take advantage because the journey of life is a long one. The journey of life is one of trials, tribulations, tests, hard work, adversity, distress, and trembling.

In this ordeal, the human being needs provisions so that the ship of life would settle at the end of the journey either in everlasting comfort or in hell and torture, God forbid. Then, oh brothers, we bid farewell to the month of Ramadan, where we found ease in what God, the Great and Almighty, had made available for us and where He opened the portals of goodness, obedience, and benefaction. We should approach the coming days with higher determination, greater persistence, stronger challenge, and stronger belief because we are in a continuous diligent battle without interruption--a battle with ourselves: 'Nor do I absolve my own self (of blame): the (human) soul is certainly prone to evil" [Koranic verse; Yusuf 12:53].

This self that loves to rest, clings to meekness, and leans toward reliance, laziness, and slothfulness along with a battle against the devil who swore that he would assault us from before us and behind us, from our right and our left. "Then will I assault them from before them and behind them, from their right and their left: Nor wilt thou find, in most of them, gratitude (for thy mercies)" [Koranic verse; al-A'raf 7:17]. He is the one who positioned himself to prevent the worshipers of God from His righteous path and a battle against the supporters of the devil and his soldiers, about whom God, the Great and Almighty, says: "Nor will they cease fighting you until they turn you back from your faith if they can" [Koranic verse; al-Baqarah 2:217]. Therefore, it is a vicious battle and a continuous battle. Its fronts are many and with different soldiers, during which the human being needs be more confident in God, the Great and Almighty, relying greatly upon God, the Great and Almighty, and seeking refuge in his Lord, the Great and Almighty, more frequently.

Consequently, oh brothers, so that we should know the value of our Lord whom we worship, the value of our Lord to whom we seek to be close, and the value of our Lord for Whose sake we fight in order to raise His word, I found that I would stop at a significant, great Hadeeth which the scholars knew for its value and to which they gave great significance. That Hadeeth, which is narrated by the Imam Muslim, with whom God is pleased and may He bless his soul, about Abu Dharr, with whom God is pleased, about the messenger of God, prayers and peace be upon him, regarding what he narrates about his Lord, he said: "God the Great and Almighty says 'and contemplate in this Hadeeth and look into its words and go deep into its meanings so that you would know your Lord to whom your are seeking to be close, the One who you are striving to meet, the One whom you are seeking to be lucky to see, the Great and Almighty. [He said:] "Oh My servants, I have forbidden oppression for Myself and have made it forbidden among you, so do not oppress one another. Oh My servants, all of you are astray except for those I have guided, so seek guidance of Me and I shall guide you. Oh My servants, all of you are hungry except for those I have fed, so seek food from Me and I shall feed you. Oh My servants, all of you are naked except for those I have clothed, so seek clothing from Me and I shall clothe you. Oh My servants, you sin by night and by day, and I forgive all sins, so seek forgiveness from Me and I shall forgive you. Oh My servants, you will not attain harming Me so as to harm Me, and will not attain benefiting Me so as to benefit Me. Oh My servants, were the first of you and the last of you, the human of you and the jinn of you, to be as pious as the most pious heart of any one man of you, that would not increase My kingdom in anything. Oh My servants, were the first of you and the last of you, the human of you and the jinn of you, to be as wicked as the most wicked heart of any one man of you, that would not decrease My kingdom in anything. Oh My servants, were the first of you and the last of you, the human of you and the jinn of you, to rise up in one place and make a request of Me, and were I to give everyone what he requested, that would not decrease what I have any more than a needle decreases the sea if put into it. Oh My servants, it is but your deeds that I reckon up for you and then recompense you for, so let him who finds good praise God, and let him who finds otherwise blame no one but himself" [Hadeeth Qudsi].

Abu Idris al-Khawlani, God bless his soul, who was one of the narrators of this Hadeeth, when he used to narrate it, he would kneel due to the greatness of this Hadeeth and because of its strong meanings, for it points out for us the effective capability of God and His great mercy and great wisdom, the Great and Almighty. See how Your Lord calls His servants, "Oh My servants." He shows love to them, and perhaps one of them would listen to His call. Then he would answer to Him, go back to Him, and listen to His words and ask forgiveness from his sins. "Oh My servants, I have forbidden oppression for Myself". The ownership is the ownership of God, the decision is the decision of God, and the entire matter is up to God, the Great and Almighty, and even so, God, the Great and Almighty has forbidden the oppression of any of His servants "nor is thy Lord ever unjust (in the least) to His Servants" [Koranic verse; Fussilat 41:46]. "We did them no wrong, but they were used to doing wrong to themselves" [Koranic verse; al-Nahl 16:118]. "Allah is never unjust in the least degree: If there is any good (done), He doubleth" [Koranic verse; al-Nisa 4:40]. This is our Lord, our Lord the Just, His judgment is the judgment of justice, and His laws are the laws of justice. This law is one that He legislated for us through his messenger, prayers and peace be upon him, and it has been descended to us in His precious book [Koran], in which [God said:] "No falsehood can approach it from before or behind it: It is sent down by One Full of Wisdom, Worthy of all Praise" [Koranic verse; Fussilat 41:42]. This Sharia is the Sharia of justice, equity, and moderation. [God Said:] "Allah commands justice, the doing of good, and liberality to kith and kin" [Koranic verse; al-Nahl 16:90]. "Oh My servants, I have forbidden oppression for Myself" [Hadeeth, Qudsi]. No one should fear God. [God said:] 'Either of short account or of any injustice'" [Koranic verse; al-Jinn 72:13].

If you come on Judgment Day as a pious good believer, a monotheist, be assured that God will not wrong you. [God said:] "Allah is never unjust in the least degree" [Koranic verse; al-Nisa 4:40]. [God said:] "I have forbidden oppression for Myself and have made it forbidden among you, so do not oppress one another" [Hadeeth Qudsi], meaning that God has forbidden oppression among people, so no one should oppress the other.

As you know dear brothers, the greatest injustice is polytheism. Injustice as defined by the scholars is to place something in an inappropriate place with infringement and abuse. The greatest is associating something as being equal to God. God has created you and you worship with Him another God and rule without the Sharia of your creator and sustainer, who gave you and protected you. This is among the greatest injustices. Polytheism is a great injustice. "I have made it forbidden among you" [Hadeeth Qudsi]. Do not commit oppression against one another. For this reason, the prophet, prayers and peace be upon him, said: "A Muslim is a brother of another Muslim, so he should not oppress him, nor should he hand him over to an oppressor" [Hadeeth].

The prophet, prayers and peace be upon him, said: "The whole of a Muslim for another Muslim is inviolable: his blood, his property, and his honor" [Hadeeth]. The prophet said: "Oppression will be darkness on the day of judgment". [Hadeeth]. What will those people who filled all the land with the darkness of oppression do? They unjustly shed the people's blood, looted their money without justification, pillaged their homes without the right to do so, and violated their honor without justification. They will come on Judgment Day as [God said:] "Bear their burdens on their backs, and evil indeed are the burdens that they bear?" [Koranic verse; al-An'am 6:31]

Oh Brothers, our religion to which we should adhere and about its correctness we should be certain is the only way that will lead us to the pleasure of Almighty God. It is the way of justice, fairness, and equality; it is a way of compassion. Those who seek justice, compassion and equity elsewhere, will only find stress and difficulty; as Almighty God said: "But whosoever turns away from My Message, verily for him is a life narrowed down, and We shall raise him up blind on the Day of Judgment" [Koranic verse; Ta-Ha 20:124]. [God said:] "Whoever works righteousness, man or woman, and has Faith, verily, to him will We give a new Life, a life that is good and pure and We will bestow on such their reward according to the best of their actions" [Koranic verse; al-Nahl 16:97].

Yes, this is our Sharia that we call and fight for and we are patient on tribulations for the sake of propagating it. This is our Sharia, which Almighty God has revealed in the precious book. What about the people whom God had guided to this great good and plentiful piety and they run East and West, North and South, seeking meekness and comfort and seeking justice [somewhere else] without the revelation of Almighty God. This is only misguidance, madness, and loss. [God said:] "Allah commands justice, the doing of good, and liberality to kith and kin" [Koranic verse; al-Nahl 16:90]. Yes, Almighty God, commanded us to call for His religion and to follow his Sharia and to be just with all, [God said:] "Now then, for that reason, call them to the Faith, and stand steadfast as thou art commanded, nor follow thou their vain desires" [Koranic verse; al-Shura 42:15].

There are two ways: straightness on the path of truth that will lead you to Almighty God, "And stand steadfast as thou art commanded, nor follow thou their vain desires" [Koranic verse; al-Shura 42:15]. This is the way to the truth and there are desires. [God said:] "Follow thou their vain desires; but say: 'I believe in the Book which Allah has sent down'" [Koranic verse; al-Shura 42:15].

Do not oppress each other. Nowadays we see the land filled with different kinds of injustices, oppression, discontent, arrogance, tyranny, and humiliation against the weak from the believers and others in a way that was never heard of before by the predecessors and the present people. Have you seen injustices that are worse and uglier than what your brothers in Gaza are experiencing? One and a half million Muslims in one closed prison! What is their guilt and their crime? Are all those terrorists? Did they all carry arms, shell, kill, or destroy? No, by God, it is a war against religion and a war against faith. There is Hezbollah, there are the righteous and good people, and there are all those who joined this religion and adhered to it and knew its value. There is Satan's party, the party of tyranny, haughtiness, and corruption, which burned the land and filled it with what? They filled it with their corruption and insolence. What is the guilt of these nursing infants and pious elderly? What is the guilt of these vulnerable women? If this injustice we are facing was from the Jews, the matter would have been easy as this is their behavior, their way, and their legacy, as God Almighty stated, "And they strive to do mischief on earth" [Koranic verse; al-Ma'idah 5:64]. A Jew is inclined to be corrupt; this is his way of life and it is in his nature. But those are criminals who affiliate with Islam. We say this so that the people of faith know who their enemies are because there is no way that we can be victorious in our battle or firmly establish Sharia unless we know the true nature of our enemies.

The cold-hearted, evil, tyrannical, despot, the modern day pharaoh from Egypt...why? Why does he sick his cronies that he has groomed over the years on these feeble people? What have they done to deserve this? What crime have these feeble ones committed? Is there a greater injustice than what is taking place against your brothers in Palestine? There isn't enough time to mention the different forms of injustice that are currently taking place, and even if we wanted to, we couldn't.

Therefore, "oh My servants, indeed I have forbidden oppression upon Myself and I have also made it forbidden amongst you, so do not oppress each other" [Hadeeth]. This is our Sharia [way of life] that we call for. But justice is not democracy or capital; it is Islam that He revealed in His book and revealed through His prophet, prayers and peace be upon him. We cannot adulate any portion or part of Sharia, but rather, "So hold thou fast to the Revelation sent down to thee; verily thou art on a Straight Way" [Koranic verse; al-Zukhruf; 43:43].

Yes, do not oppress each other. "Oh My servants, all of you are astray except for those I have guided, so seek guidance of Me and I shall guide you" [Hadeeth]. Therefore, there is only one source of guidance, one path to guidance, and he who seeks guidance must seek God Almighty. He who seeks guidance must seek guidance of God Almighty. Anything other than that straying from the right path, then, "Show us the straight way, the way of those on whom Thou hast bestowed Thy Grace, those whose (portion) is not wrath, and who go not astray" [Koranic verse; al-Fatiha; 1:6; 1:7].

Yes, "oh My servants, all of you are astray except for those I have guided" [Hadeeth]. Therefore, guidance to the right path can only be achieved by adhering to God Almighty's Sharia. Guidance to the right path can only be achieved through God's light, therefore enabling it to become part of the monotheist Ummah. God will guide those who He wants by opening their hearts to Islam. He will make the hearts of those who he wants go astray, cold and closed off. Yes, even if the world's riches were bestowed upon him, he would still feel an endless sense of distress. Why you ask? It is because this heart, or this small piece of it, will not feel peaceful or content until it returns to its origin, the way it was created by God, as the belief in God Almighty and the adherence to His Sharia, monotheism, the love, and the fear of God.

As for a person who disperses his heart [beliefs] left and right by following more than one ideology--sometimes worshipping a rock, sometimes worshipping a tree, sometimes worshipping air, and other times worshipping laws gone astray – he will never achieve happiness. "But whosoever turns away from My Message, verily for him is a life narrowed down, and We shall raise him up blind on the Day of Judgment" [Koranic verse; Taha; 20:124].

Oh you Muslims who have violated God's Sharia and the light of his prophet and left it all behind you as you search for worldly pleasures, happiness, and contentment in the ideological trash of the grandchildren of pigs and monkeys, return to your God and ask for His guidance, "so seek guidance of Me and I shall guide you" [Hadeeth]. Yes, oh My servants, all of you are astray except for those I have guided" [Hadeeth].

"Oh My servants, all of you are hungry except for those I have fed, so seek food from Me and I shall feed you. Oh My servants, all of you are naked except for those I have clothed, so seek clothing from Me and I shall clothe you" [Hadeeth]. This is our God; all the wealth and blessings of this world and the hereafter are in His hands. He who desires a portion of worldly pleasure, whether big or small, whether it is food or clothes or good health and well-being, must seek God Almighty for guidance.

God does not consider anything too great that he is asked for by his worshippers. Ibrahim, peace be upon him, amidst the worst of his ordeal, was alone, a stranger among his people, yet still declared among them, loud and clear, as every Muslim should, his devotion to monotheism, his confidence in God Almighty, his reliance on Him, and the realization of His worth by saying to his people, "What worship ye? They said: 'We worship idols, and we remain constantly in attendance on them.' He said: 'Do they listen to you when ye call (on them)' Or do you good or harm?" [Koranic verses; al-Shu'ara; 26:71-73]. He also says, "Who created me, and it is He Who guides me, Who gives me food and drink, And when I am ill, it is He Who cures me; Who will cause me to die, and then to life (again); And who, I hope, will forgive me my faults on the day of Judgment. O my Lord! Bestow wisdom on me, and join me with the righteous" [Koranic verses; al-Shu'ara; 26:78-83].

Yes, this how the confidence and certainty of Muslims in their God Almighty should be. [They must] rely on Him and trust in Him, take refuge in Him, beseech Him, and have confidence in what He, the Almighty, has [revealed] and place their faith in his hands and be confident of what He, the Almighty has to offer.

"Oh My servants, all of you are hungry except for those I have fed." Then, all the sustenance of heaven and earth are from the hands of God Almighty, thus, with all that God has to offer, it is not possible for the springs of terror to dry up. Why? It is because all sustenance is from the hands of God Almighty. This calling for which all these summits have been held and upon which the devils of humanity and jinn converge, has been discussed previously. They said, "Do not spend on the followers of the messenger of God so that they may disperse." This is an old calling and not a new one, seeking to dry out the springs of terror. To that we say that the hands of God control the wealth of the heavens and earth so you can cut off what you wish, dry up what you want, and forbid what you desire, but if God Almighty sought anything, all He has to do is command it to be, and so it shall be. If God was to command something to be, the heavens and earth would not be able stop Him.

My brothers, this sense of confidence must be planted in our hearts. Oh people of the mujahideen, across the world from East to West, you must be fully confident in your Lord, in your faith, and in your lives. Seek substance from God Almighty. Seek more from what He has to offer, for your Lord is calling you to seek him, to ask him, and to supplicate to him. "All of you are hungry except for those I have fed, so seek food from Me and I shall feed you." So much so that some of the elders at that time would ask God for the salt of their foods or the food of their cattle. This is how our ancestors were, so we should not be embarrassed to ask God Almighty for anything since this action pushes us toward Him, the Almighty.

"Oh My servants, all of you are hungry except for those I have fed, so seek food from Me and I shall feed you. Oh My servants, all of you are naked except for those I have clothed, so seek clothing from Me and I shall clothe you. Oh My servants, you sin by night and by day, and I forgive all sins, so seek forgiveness from Me and I shall forgive you." All human beings are sinners and among the best of them are those who repent. God Almighty has informed us that there are no human beings who do not make mistakes by day or night. God Almighty opens His hands at night to forgive those who have sinned by day and extends His arms by day to forgive those who have sinned at night. There is nothing that will bring more trials and tribulations than to have sinned, whether it is against a person or against a community. Also, "Such is the chastisement of thy Lord when He chastises communities in the midst of their wrong: grievous, indeed, and severe is His chastisement" [Koranic verse; al-Hud; 11:102]. "But the Unbelievers,- never will disaster cease to seize them for their (ill) deeds, or to settle close to their homes, until the promise of Allah come to pass, for, verily, Allah will not fail in His promise" [Koranic verse; al-Ra'd; 13:31].

This is true for us my brothers, and especially true of the mujahideen, for you are in biggest need of repentance to the Almighty God and need to seek His forgiveness. Why? It is because your sins pose the greatest obstacle between yourselves and victory. God will not grant victory to people who disobey him. This is why the people of faith, before they ask God to grant them victory over the nonbelievers, must supplicate: "Our Lord! Condemn us not if we forget or fall into error; our Lord! Lay not on us a burden Like that which Thou didst lay on those before us; Our Lord! Lay not on us a burden greater than we have strength to bear. Blot out our sins, and grant us forgiveness. Have mercy on us. Thou art our Protector; Help us against those who stand against faith" [Koranic verse; al-Baqarah; 2:286].

Then, we are in need of increasing our seeking of forgiveness and repentance to God Almighty. Seek this true and honest repentance in which what is hidden and what is displayed become one: that of which you speak and what you keep secret. This is true because God does not judge you based on your appearance or your color, however He does judge you by looking into your hearts and seeing the kind of honesty, truthfulness, and monotheism. He also judges you by your actions and how they compare to what the prophet, peace and blessings upon him, has prescribed.

So, if we are looking for victory and triumph, we must seek a true repentance and constant supplications of forgiveness from God Almighty. "How many of the prophets fought (in Allah's way), and with them (fought) Large bands of godly men? but they never lost heart if they met with disaster in Allah's way, nor did they weaken (in will) nor give in. And Allah Loves those who are firm and steadfast. All that they said was: 'Our Lord! Forgive us our sins and anything We may have done that transgressed our duty: Establish our feet firmly, and help us against those that resist Faith'" [Koranic verse; al-Imran; 3:146-147].

My brothers, we need to remind ourselves of this great issue for, by God, this is one of the greatest paths to victory, glory, and triumph, as long as the people of the mujahideen are close to God Almighty, seeking repentance from God Almighty, and seeking forgiveness from God Almighty, and the more they seek his intervention, the closer they will be to victory.

"Oh My servants, you sin by night and by day, and I forgive all sins, so seek forgiveness from Me and I shall forgive you."

God almighty says, "Oh My servants, you sin by night and by day, and I forgive all sins, so seek forgiveness from Me and I shall forgive you." God Almighty has no need for us, for our guidance, for our obedience, for our welfare, or for our sincerity. God Almighty commands what is in the heavens and earth: "O ye men! It is ye that have need of Allah. But Allah is the One Free of all wants, worthy of all praise" [Koranic verse; Fatir; 35:15]. This is so no man can become arrogant with faith or not be impressed by his good deeds, those to whom God has guided him. Almighty God explained this fact to us in a clear and explicit way which everyone can understand: "Oh My servants, you will not attain harming Me so as to harm Me and will not attain benefiting Me so as to benefit Me' [Hadeeth], and 'Let not those grieve thee who rush headlong into Unbelief; Not the least harm will they do to Allah" [Koranic verses, al-Imran, 3:175] and "Those who reject Allah, hinder (men) from the Path of Allah, and resist the Messenger, after Guidance has been clearly shown to them, will not injure Allah in the least, but He will make their deeds of no effect" [Koranic verse; Mohammed 47:32]. Yes, then God said; "Oh My servants, were the first of you and the last of you, the human of you and the jinn of you to be as pious as the most pious heart of any one man of you, that would not increase My kingdom in anything" [Hadeeth].

Almighty God does not need us or need our prayers, jihad, strife, binding, or supplication, and that is why God says at the end of the Hadeeth [Qudsi]: "It is but your deeds that I reckon up for you and then recompense you for, so let him who finds good praise Allah and let him who finds otherwise to blame no one but himself." So, a Muslim must not be conceited or deceived by his good deeds that God enabled him to comply with, for it is a blessing and gratitude to worship Almighty God and to rejoice for what God granted him and increase his appreciation.

"Oh My servants, were the first of you and the last of you, the human of you and the jinn of you to rise up in one place and make a request of Me, and were I to give everyone what he requested, that would not decrease what I have. Oh My servants, it is but your deeds that I reckon up for you and then recompense you for, so let him who finds good praise Allah and let him who finds otherwise to blame no one but himself" [Hadeeth]. God owns the universe.

In saying; "Oh My servants, were the first of you and the last of you, the human of you and the jinn of you to rise up in one place and make a request of Me, and were I to give everyone what he requested, that would not decrease what I have, any more that a needle decreases the sea if put into it" [Hadeeth], scholars agreed that these words meant that nothing could reduce God's might, but then how could words decrease God's might and grant when [God] create the universe How could anything decrease God's might? Almighty God has all the universe of day and night with everything in between.

God said: "They are the ones who say, 'Spend nothing on those who are with Allah's Messenger, to the end that they may disperse (and quit Medina).' But to Allah belong the treasures of the heavens and the earth; but the Hypocrites understand not" [Koranic verses, al-Munafiqun, 63:7]. The meaning of these words, dear brothers, shows how merciful and gracious Almighty God is.

The words of "Oh my Servants who have transgressed against their souls! Despair not of the Mercy of Allah. For Allah forgives all sins: for He is Oft-Forgiving, Most Merciful" [Koranic verses, al-Zumar, 39:53] show the immense and vast power of God to "the first of you and the last of you" and He says: "that would not decrease what I have, any more that a needle decreases the sea if put into it" [Hadeeth]. Therefore, the power of Almighty God is immense as He adds: "and will not attain benefiting Me so as to benefit Me" [Hadeeth]. This shows that God does not need His believers at all, but when these words take root in the heart of a believer and when he is aware of them and knows that he worships a God whose description is such, God Almighty then calls him to increase his adherence to that which came from Him, and an awareness that every single word that comes from the book of God is truth and that every ruling made according to the Sharia of God is truth from God Almighty. All this calls toward increasing efforts, increasing giving, and sacrificing for the sake of the religion of God Almighty. It should call him toward having more patience and endurance in the face of everything he faces. It is important that we know, oh brothers, that the path on the road to enabling the religion of God Almighty is one of the trials, tribulations, ordeals, trembling, and adversity, and after all this comes the victory from God Almighty. Yes, God Almighty promised to empower his servants if they support Him. God Almighty said: "If ye will aid (the cause of) Allah, He will aid you and plant your feet firmly" [Koranic verse; Mohammed 47:7].

However, this victory does not take place through laziness, sluggishness, and wishes. Rather, it comes through effort, sacrifice, suffering, and patience. Then, the victory of

God Almighty takes place. God Almighty said: "Or do ye think that ye shall enter the Garden (of bliss) without such (trials) as came to those who passed away before you? they encountered suffering and adversity, and were so shaken in spirit that even the Messenger and those of faith who were with him cried: When (will come) the help of Allah?" [Koranic verse; al-Baqarah 2:214]

The prophet, to whom the revelations of God were sent, says: "When (will come) the help of Allah?" [Koranic verse; al-Baqarah 2:214]. Why does he say it? He says it due to the significant difficulty, shock, and adversity that he and his companions encountered. "Verily, the help of Allah is (always) near!" [Koranic verse; al-Baqarah 2:214]

Therefore, oh mujahideen, whether in Afghanistan, Iraq, Somalia, Algeria, or Chechnya, you should adapt to this truth. You must resign yourself to the fact that it is not an easy task to empower God Almighty. It cannot be achieved through wishes, imaginations, sluggishness, and laziness. You must adapt to consecutive trials and hardships. Later, once your ranks are purified, when the evil is distinguished from the good and the people who are truly sincere in their faith are distinguished from the hypocrites who infiltrate the ranks of the mujahideen claiming they are with them, after all this, the victory of God, to Whom belongs might and majesty, will come to us. [The victory will take place] when God sees that we are qualified to receive that great trust which is to rule per the law of God Almighty, be just with people, and pass on His rules.

Do not let the hardships you are experiencing make you give up! Do not think that the path of jihad, victory, and strengthening is easy. Rather, it is the path of shortage in money, souls, and rewards. Today, you find the field of jihad full of heroes, commanders, and experts. Suddenly, you find that the battle has crushed them and that they are gone. This is a trial of which God, to Whom belongs might and majesty, informed us. "Be sure we shall test you with something of fear and hunger, some loss in goods or lives or the fruits (of your toil), but give glad tidings to those who patiently persevere" [Koranic verse; al-Baqarah 2:155].

Oh God, bring victory to your servants, the mujahideen, anywhere they might be, and support that victory!

Oh God, support your servants, the faithful mujahideen, assist them, and grant them a manifest victory!

Oh God, support your servants, the mujahideen, anywhere they might be!

Oh God, grant them the blessings of the heavens and the earth!

Oh God, the Lord of mankind, overwhelm them with patience!

Oh God, bring them victory in Afghanistan!

Oh God, bring them strength with which you are pleased!

Oh God, bring them back their control that is better, stronger, larger, and quicker than in the past!

Oh God, bring victory to your servants, the mujahideen, in Iraq!

Oh God, remove their hardship!

Oh God, remove their loss!

Oh God, the Lord of mankind, remove their trial!

Oh God, the Strong and able to enforce His will, enrage your enemies through them!

Oh God, bring victory to your servants, the mujahideen, in Somalia!

Oh God, Lord of mankind, bring them strength with which you are pleased, through which you empower your servants and humiliate your enemies!

Oh God, bring victory to your servants, the mujahideen, in Algeria!

Oh God, remove their estrangement!

Oh God, remove their estrangement!

Oh God, grant them a manifest victory!

Oh God, Who defends Muslims, defend them!

Oh God, grant them the blessings of the heavens and the earth!

Oh God, Lord of mankind, make them only need you instead of needing others!

Oh God, bring victory to the mujahideen in Chechnya and Palestine!

Oh God, grant them a manifest victory!

Oh God, Lord of mankind, humiliate your enemies and theirs!

Oh God, Lord of mankind, release our brothers who are imprisoned in prisons of the Jews and Christians, the atheists and apostates, rejectionists and Buddhists, and Hindus!

Oh God, make every hardship a relief for them, and find an exit to every difficulty and grant them means of living from places they do not count on!

Oh God, Lord of mankind, bring them back to their families safe and sound, with spoils and reward!

Oh God, Lord of mankind, replace them with good deeds to their families, and substitute every mujahid and martyr with good deeds!

You are the one Who is close to us, Who listens, and Who responds to our supplications!

Oh God, take on your enemies, the enemies of Islam!

Oh God, take on those who fight your religion!

Oh God, take on those who fight your religion, humiliate your servants, and reject your Sharia!

Oh God, count them by the numbers, kill and remove them, and do not leave any one of them!

Oh God, take on the US and its allies!

Oh God, take on the US and its allies!

Oh God, Lord of mankind, pour agony on them like rain and inflict them with calamities from every corner!

Oh God, Lord of mankind, humiliate them in their land, overwhelm them by each other, by their people, by your soldiers, and by your faithful servants!

Oh God, take revenge on them for your oppressed servants!

Oh God, take revenge on them for your conquered servants!

Oh God, Lord of mankind, humiliate Bush and his party!

Oh, God, degrade and defy him!

Oh God, Lord of mankind, defy him!

Oh God, make him live a day like the day the pharaoh, Haman, and

Qarun experienced, making him a warning example to those who fear God and a remedy for the hearts of the believers!

You are the one Who is close to us, listens, and responds to our supplications!

May God's prayers and peace be upon the best of your creatures, Mohammed, and upon his family and all of his Companions!

al-Qaeda DVD Listing

(the below list represents what was available as of 30 Apr. 2008, check http://www.intelcenter.com for the most up-to-date listing and to order)

- Vol. 1 - The Nineteen Martyrs (English Edition)
- Vol. 2 - American Hell in Afghanistan & Iraq (English Edition)
- Vol. 3 - Badr al-Riyadh
- Vol. 4 - Wills of the NY & Washington Battle Martyrs (English Edition)
- Vol. 5 - Osama bin Laden Address to Americans & Muslims in Iraq, Aug 2003 (English Edition)
- Vol. 6 - Martyrs of Confrontations on the Arabian Peninsula (Arabic Edition)
- Vol. 7 - The Destruction of the Destroyer USS Cole: Preparation & al-Jihad (English Edition)
- Vol. 8 - The Destruction of the Destroyer USS Cole: Ummah, Cause, Migration (English Edition)
- Vol. 9 - War of the Oppressed: Part 1
- Vol. 10 - War of the Oppressed: Part 2
- Vol. 11 - London Claim - Mohammed Sidique Khan Martyrdom Video (English Edition)
- Vol. 12 - Azzam al-Amriki Statement on 11 Sep. 2005 (English Edition)
- Vol. 13 - Ayman al-Zawahiri on 23 Oct. 2005 (Arabic Edition)
- Vol. 14 - Ayman al-Zawahiri on 29 Nov. 2004 (Arabic Edition)
- Vol. 15 - Ayman al-Zawahiri on 19 Sep. 2005 (English Edition)
- Vol. 16 - Ayman al-Zawahiri: Impediments to Jihad (English Edition)
- Vol. 17 - Ayman al-Zawahiri: Letter to the Americans (English Edition)
- Vol. 18 - Ayman al-Zawahiri on 17 Jun. 2005 (Arabic Edition)
- Vol. 19 - Ayman al-Zawahiri: The Victory of Islam in Iraq (English Edition)
- Vol. 20 - American Inferno in Khorasan Series: Set 1
- Vol. 21 - Osama bin Laden 19 Jan. 2006 Audio Address Video with English Subtitles
- Vol. 22 - Osama bin Laden 19 Jan. 2006 Audio Address Video with English Voiceover
- Vol. 23 - Ayman al-Zawahiri: Bajawr Massacre (English Edition)
- Vol. 24 - Ayman al-Zawahiri on 4 Mar. 2006
- Vol. 25 - Ayman al-Zawahiri: From Tora Bora to Iraq
- Vol. 26 - Osama bin Laden: Oh People of Islam (English Subtitles)
- Vol. 27 - Ayman al-Zawahiri: A Letter to the People of Pakistan
- Vol. 28 - Osama bin Laden: Thoughts Over al-Aqsa Intifadah
- Vol. 29 - Osama bin Laden: A Testimony to the Truth (English Subtitles)
- Vol. 30 - Ayman al-Zawahiri on 9 Jun. 2006 (English Subtitles)
- Vol. 31 - Abu Yahya al-Libi on 17 Jun. 2006 (English Subtitles)
- Vol. 32 - al-Khobar 29 May 2004 Hostage Taking
- Vol. 33 - American Crimes in Kabul
- Vol. 34 - *Not Yet Listed*
- Vol. 35 - Abu Yahya al-Libi on 11 May 2006
- Vol. 36 - Omar al-Faruq on 27 Feb. 2006
- Vol. 37 - Ayman al-Zawahiri: Elegizing Abu Musab al-Zarqawi (English Subtitles)
- Vol. 38 - Osama bin Laden: Elegizing Abu Musab al-Zarqawi
- Vol. 39 - Osama bin Laden: To the Ummah in General and to the Mujahideen in Iraq and Somalia in Particular
- Vol. 40 - Wills of the Knights of the London Raid (Part II) (English Subtitles)
- Vol. 41 - Ayman al-Zawahiri: The Zionist Crusader's Aggression on Gaza & Lebanon
- Vol. 42 - Abu Yahya al-Libi: Light & Fire in Eulogizing the Martyr Abu Musab al-Zarqawi (English Subtitles)
- Vol. 43 - Abu Jihad al-Masri: Communique from those adhering to the covenant in the Egyptian Islamic Group (English Subtitles)
- Vol. 44 - Azzam al-Amriki: An Invitation to Islam (English Subtitles)
- Vol. 45 - Knowledge is for Acting Upon: The Manhattan Raid (English Subtitles)
- Vol. 46 - Hot Issues with Ayman al-Zawahiri

(English Subtitles)
• Vol. 47 - Abu Yahya al-Libi: Jihadi Poems
• Vol. 48 - Ayman al-Zawahiri: Bush, the Vatican's Pope, Darfur and the Crusaders (English Subtitles)
• Vol. 49 - Abu Yahya al-Libi: Combat, Not Compromise (English Subtitles)
• Vol. 50 - American Incinerator in Khorasan Series: Abu Nasir al-Qahtani: 1st Op After Bagram Escape
• Vol. 51 - American Incinerator in Khorasan Series: Set 2
• Vol. 52 - American Incinerator in Khorasan Series: Abu Nasir al-Qahtani: Khost Raid
• Vol. 53 - Ayman al-Zawahiri: Realities of the Conflict Between Islam and Unbelief (English Subtitles)
• Vol. 54 - Interview with Mujahid Commander Mullah Dadullah (English Subtitles)
• Vol. 55 - Ayman al-Zawahiri: Congratulation on the Eid to the Ummah of Tawhid (English Subtitles)
• Vol. 56 - Ayman al-Zawahiri: Rise and Support Your Brothers in Somalia
• Vol. 57 - Ayman al-Zawahiri: The Correct Equation (English Subtitles)
• Vol. 58 - Abu Yahya al-Libi: And the Crusade Continues... The AIDs Children in Libya (English Subtitles)
• Vol. 59 - Abu Yahya al-Libi: The Adha Holiday Sermon 1427H
• Vol. 60 - Ayman al-Zawahiri: Tremendous Lessons and Events in the Year 1427H (English Subtitles)
• Vol. 61 - Holocaust of the Americans In the Land of Khorasan, The Islamic Emirate: Capture of an American Post, Arghandab (English Voiceover)
• Vol. 62 - Holocaust of the Americans In the Land of Khorasan, The Islamic Emirate: Capture of an American Post, Arghandab (English Subtitles)
• Vol. 63 - Holocaust of the Americans In the Land of Khorasan, The Islamic Emirate: Martyrdom Operation Against an American Convoy in Argon Area
• Vol. 64 - Ayman al-Zawahiri: Palestine is Our Concern and the Concern of Every Muslim (English Subtitles)
• Vol. 65 - Abu Yahya al-Libi: Iraq: Between Indications of Victory and Conspiratorial Intrigues (English Subtitles)
• Vol. 66 - Abu Yahya al-Libi: To the Army of Difficulty in Somalia (English Subtitles)
• Vol. 67 - Ayman al-Zawahiri Interview Sep 2002 (English Subtitles)
• Vol. 68 - Holocaust of the Americans In the Land of Khorasan, The Islamic Emirate: Martyrdom Operation Against an American Convoy in the State of Baktika
• Vol. 69 - Osama bin Laden 29 Oct. 2004 Address to Americans (English Subtitles)
• Vol. 70 - Osama bin Laden Tarnak Farms Address and Last Wills of 9-11 Hijackers Ziad Jarrah and Mohammed Atta
• Vol. 71 - Osama bin Laden: The Speech of Ka'b bin Malik on Tabuk Raid
• Vol. 72 - Interview with Sheikh Abu Laith, One of the Leaders of Qaeda al-Jihad in Khorasan
• Vol. 73 - Abu Yahya al-Libi: Palestine, an Alarming Scream and a Warning Cry
• Vol. 74 - Interview with Sheikh Ayman al-Zawahiri, April/May 2007 (English Subtitles)
• Vol. 75 - Ayman al-Zawahiri: Elegizing the Commander of the Martyrdom-Seekers Mullah Dadullah (English Subtitles)
• Vol. 76 - Abu Laith al-Libi: Confronting the War of Prisons (English Subtitles)
• Vol. 77 - Interview with Sheikh Mustafa Abu al-Yazid "Sheikh Saeed" (English Subtitles)
• Vol. 78 - Legitimate Demands, A Message from the Mujahid Brother Adam Yahiye Gadahn (Azzam)
• Vol. 79 - Abu Yahya al-Libi: The Tawheed of Saud... and the True Tawheed (English Subtitles)
• Vol. 80 - Holocaust of the Americans in the Land of Khorasan, The Islamic Emirate: Attack on an Apostate Base, Dabgay, Khost Province (English Subtitles)
• Vol. 81 - Holocaust of the Americans in the Land of Khorasan, The Islamic Emirate: Firing of BM Rockets on a Base in Lwara, Hitting it Directly
• Vol. 82 - Abu Yahya al-Libi: Elegy for Mullah Dadullah

• Vol. 83 - Holocaust of the Americans in the Land of Khorasan, The Islamic Emirate: Ambush of a Police Motorcade on the Kabul-Kandahar Highway in Zabul Province (English Subtitles)
• Vol. 84 - Ayman al-Zawahiri: Forty Years Since the Fall of Jerusalem
• Vol. 85 - Ayman al-Zawahiri: The Advice of One Concerned (English Subtitles)
• Vol. 86 - Ayman al-Zawahiri: Malicious Britain and Its Indian Slaves
• Vol. 87 - Ayman al-Zawahiri: The Aggression Against Lal Masjid (English Subtitles)
• Vol. 88 - Winds of Paradise, Part 1
• Vol. 89 - Abu Yahya al-Libi: The Masters of the Martyrs
• Vol. 90 - Abu Yahya al-Libi: The Masters of the Martyrs (Urdu Subtitles)
• Vol. 91 - The Will of the Martyr Hafiz Usman (English Subtitles)
• Vol. 92 - The Will of the Martyr Hafiz Usman (Arabic Subtitles)
• Vol. 93 - The Will of the Martyr Hafiz Usman (Urdu Subtitles)
• Vol. 94 - The Two Sheikhs on the Mountains of Islam
• Vol. 95 - The Wind of Paradise, Part 1 (English Subtitles)
• Vol. 96 - Osama bin Laden: The Solution (English Subtitles)
• Vol. 97 - Abu Yahya al-Libi: Dots on the Letters (English Subtitles)
• Vol. 98 - The Wills of the Heroes of the Raids on New York and Washington, The Will of the Martyr Abu Musab Walid al-Shehri (English Subtitles)
• Vol. 99 - The Power of Truth (English Subtitles)
• Vol. 100 - Osama bin Laden: Come to Jihad (English Subtitles)
• Vol. 101 - Osama bin Laden: Come to Jihad (Pashto Voiceover & English Subtitles)
• Vol. 102 - Osama bin Laden: Come to Jihad (Urdu Voiceover & English Subtitles)
• Vol. 103: Mustafa Abu al-Yazid: The Truth About Reliance on God (Pashto Version)
• Vol. 104: Osama bin Laden: A Message to our People in Iraq (English Subtitles)
• Vol. 106: Osama bin Laden: Eid al-Fitr Address 1999 (Urdu Subtitles)
• Vol. 107: Interview with Mansour Dadullah (Arabic Subtitles)
• Vol. 108: Interview with Mansour Dadullah (Pashto Subtitles)
• Vol. 109: Unity of Ranks
• Vol. 110: Abu Yahya al-Libi: The Closing Statement for the Religious Training that was Held at One of the Mujahideen Centers
• Vol. 111: Osama bin Laden: To the European Peoples (English Subtitles)
• Vol. 112: Osama bin Laden: To the European Peoples (German Subtitles)
• Vol. 113: Osama bin Laden: To the European Peoples (Pashto Voiceover)
• Vol. 114: Ayman al-Zawahiri: Annapolis - The Betrayal
• Vol. 115: Ayman al-Zawahiri: A Review of Events, as-Sahab's Fourth Interview with Sheikh Ayman al-Zawahiri
• Vol. 116: Abu Yahya al-Libi: Going Forth
• Vol. 117: Osama bin Laden: The Way to Foil Conspiracies on Iraq and the Islamic State
• Vol. 118: Adam Gadahn: An Invitation to Reflection and Repentance (English)
• Vol. 119: Abu Yahya al-Libi: An Eid al-Adha Speech
• Vol. 120: The Wind of Paradise, Part 2
• Vol. 121: Mustafa Abu al-Yazid: Light and Fire: An Announcement to the Ummah (English Subtitles)
• Vol. 122: Ayman al-Zawahiri: An Elegy to the Martyred Commander Abu Laith al-Libi (English Subtitles)
• Vol. 123: Abu Yahya al-Libi: The Companion
• Vol. 124: Mustafa Abu al-Yazid: They Lied, Now is the Time to Fight
• Vol. 125: Abu Yahya al-Libi: I am Not a Deceiver, Nor Will I Allow Someone to Deceive Me
• Vol. 126: Osama bin Laden: May Our Mothers be Bereaved of Us if We Fail to Help Our Prophet (PBUH) (English Subtitles)
• Vol. 127: Osama bin Laden: The Way for the Salvation of Palestine
• Vol. 128: Ayman al-Zawahiri: Rise to Support Our Kinfolk in Gaza